PRODUCTION & OPERATIONS MANAGEMENT

AUTHOR

Mayank Kumar Agrawal
Lecturer
Department of Mechanical Engineering
Technical College,
Dayalbagh Educational Institute (Deemed to be University),
Agra, 282005, Uttar Pradesh, India

REVIEWER

Prof. C. P. Dewangan
Department of Industrial and Production
Engineering
Guru Ghasidas Vishwavidyalaya,
Bilaspur, 495009, Chhattisgarh, India

KHANNA BOOK PUBLISHING CO. (P) Ltd.
4C/4344, Ansari Road, Darya Ganj,
New Delhi, India

All India Council for Technical Education
Nelson Mandela Marg, Vasant Kunj,
New Delhi, India

ISBN: 978-93-5538-779-0

Book Code: DIP505EN

Production And Operations Management

by Dr. Mayank Kumar Agrawal

First Edition: 2026

Originally Published by:

All India Council for Technical Education, Ministry of Education, GOI

Exclusively Published & Distributed by:

Khanna Book Publishing Co. (P) Ltd.

4C/4344, Ansari Road, Daryaganj,
New Delhi (INDIA)

Visit us at: **www.khannabooks.com**

Write us at: **contact@khannabooks.com**

To view complete list of books, Please scan the QR Code: KPH

Printed in Bharat.

प्रो. टी. जी. सीताराम
अध्यक्ष
Prof. T. G. Sitharam
Chairman

सत्यमेव जयते

आज़ादी का अमृत महोत्सव

अखिल भारतीय तकनीकी शिक्षा परिषद्
(भारत सरकार का एक सांविधिक निकाय)
(शिक्षा मंत्रालय, भारत सरकार)
नेल्सन मंडेला मार्ग, वसंत कुंज, नई दिल्ली–110070
दूरभाष : 011–26131498
ई–मेल : chairman@aicte-india.org

ALL INDIA COUNCIL FOR TECHNICAL EDUCATION
(A STATUTORY BODY OF THE GOVT. OF INDIA)
(Ministry of Education, Govt. of India)
Nelson Mandela Marg, Vasant Kunj, New Delhi-110070
Phone : 011-26131498
E-mail : chairman@aicte-india.org

FOREWORD

Engineers are the backbone of any modern society. They are the ones responsible for the marvels as well as the improved quality of life across the world. Engineers have driven humanity towards greater heights in a more evolved and unprecedented manner.

The All India Council for Technical Education (AICTE), have spared no efforts towards the strengthening of the technical education in the country. AICTE is always committed towards promoting quality Technical Education to make India a modern developed nation emphasizing on the overall welfare of mankind.

An array of initiatives has been taken by AICTE in last decade which have been accelerated now by the National Education Policy (NEP) 2020. The implementation of NEP under the visionary leadership of Hon'ble Prime Minister of India envisages the provision for education in regional languages to all, thereby ensuring that every graduate becomes competent enough and is in a position to contribute towards the national growth and development through innovation & entrepreneurship.

One of the spheres where AICTE had been relentlessly working since past couple of years is providing high quality original technical contents at Under Graduate & Diploma level prepared and translated by eminent educators in various Indian languages to its aspirants. For students pursuing 2nd year of their Engineering education, AICTE has identified 87 books, which shall be translated into 12 Indian languages - Hindi, Tamil, Gujarati, Odia, Bengali, Kannada, Urdu, Punjabi, Telugu, Marathi, Assamese & Malayalam. In addition to the English medium, books in different Indian Languages are going to support the students to understand the concepts in their respective mother tongue.

On behalf of AICTE, I express sincere gratitude to all distinguished authors, reviewers and translators from the renowned institutions of high repute for their admirable contribution in a record span of time.

AICTE is confident that these outcomes based original contents shall help aspirants to master the subject with comprehension and greater ease.

T. G. Sitharam

(Prof. T. G. Sitharam)

ACKNOWLEGMENT

The author is grateful to the authorities of AICTE, particularly Prof. (Dr.) T.G. Sitharam, Chairman, Dr. Abhay Jere, Vice-Chairman, Prof. Rajive Kumar, Member-Secretary, Dr. Sunil Luthra, Director and Reena Sharma, Hindi Officer Training and Learning Bureau for their decision to publish this book on Production and Operations Management. I sincerely acknowledge the valuable contributions of the reviewer of the book Prof. C. P. Dewangan, Associate professor of S. Guru Ghasidas Vishwavidyalaya, Bilaspur.

This book is an outcome of various suggestions of AICTE members, experts, and authors who shared their opinion and thoughts to further develop engineering education in our country. Acknowledgements are due to the contributors and different workers in this field whose published books, review articles, papers, photographs, footnotes, references and other valuable information enriched the final draft of the book.

The author is deeply impressed by the academic prowess of the Director of the Dayalbagh Educational Institute (Deemed to be University), Dayalbagh, Agra, Prof. C. Patvardhan who has been a source of great inspiration in motivating the author towards academic pursuits. The Registrar, Prof. Anand Mohan is a workaholic himself and inspires not to be deterred by seemingly unsurmountable obstacles. The author is indebted to the Principal, D.E.I. Technical College, Er. Vijay Prakash Malhotra for his insightful comments and pieces of advice from time to time which helped take the book to its final shape. Sh. G.P. Misra, Head, Mechanical Engineering deserves thanks for his support in the department. Prof. D.K. Chaturvedi, Head/Coordinator Dept of Footwear Technology, DEI, with whom the author worked on several projects, deserves special mention for his support and motivation. My parents Sh. Rajendra Agrawal and mother Smt. Madhu Agrawal deserve special appreciation for their patience and understanding when I missed spending my precious time with them. My wife Mrs. Drishti and daughter Surat deserve special thanks for their support and encouragement and patience for times when I could not spare time for them in completion of this monumental task.

Mayank Kumar Agrawal

PREFACE

The book titled "Production & Operations management" is an outcome of the rich experience of our teaching of basic and advanced courses of Production Technology and Operational Management. Keeping in mind the purpose of wide coverage as well as to provide essential supplementary information, we have included the topics recommended by AICTE, in a very systematic and orderly manner throughout the book. Efforts have been made to explain the fundamental concepts of the subject in the simplest possible way.

During the process of preparation of the book, we have considered the various standard text books and accordingly we have developed sections and a variety of questions like multiple choice, short and long answer, numerical problems and supplementary material. While preparing the different sections emphasis has also been laid on examples, supplementary material. The book covers all types of medium and advanced level problems and these have been presented in a very logical and systematic manner. The gradations of those problems have been tested over many years of teaching to a wide variety of students.

Apart from illustrations and examples as required, we have enriched the book with numerous solved problems in every unit for proper understanding of the related topics. Under the common title "Production and operation Management" there is a set of five chapters covering different aspects of operations used in industry.

This book covers all the basics of supply chain management, manufacturing planning and control systems, purchasing, physical distribution, process, quality and its management using Lean, Six Sigma and Total Quality Management. The material, examples, questions, and problems lead the student logically through the text. The writing style is simple and user-friendly-both instructors and students who would use the book will attest to this.

We sincerely hope that the book will inspire the students to learn and discuss the ideas behind basic principles of line balancing and will surely contribute to the development of a solid foundation of the subject. We would be thankful to all beneficial comments and suggestions which will contribute to the improvement of the future editions of the book. It gives us immense pleasure to place this book in the hands of the teachers and students. It was indeed a big pleasure to work on different aspects covered in the book.

Finally, the author sincerely hopes that the book will meet the needs of the readers. All the suggestions to improve the text of the book will always be welcome.

Mayank Kumar Agrawal

About the Contributor

Mr. Mayank Kumar Agrawal, working as a Lecturer in Department of Mechanical Engineering, In Technical College, Dayalbagh Educational Institute (Deemed to be University), Dayalbagh, Agra 282005 (UP), India Since 2013. He did his B.Sc. Engineering and M. Tech. from D.E.I. in May 2010 and 2018 respectively. He was awarded with the most prestigious founder medal for best student among all under graduate award in 2010 and Director's Medal for securing the highest marks in Comparative Study of Religion and Cultural Education in 2010. He represented the university at National level inter University Hindi Debate competition in A.M.U. (2009), Delhi University (2008) and National Co-Operative Society (2007) and won prizes for the Best Team and Best Speaker. Through placement drive he was recruited in Maruti Suzuki India limited, Punj Lloyd and L&T. He joined Maruti Suzuki India Limited in June 2010 and worked there as an Assistant Manager for three years. In this duration he worked in Production planning and Control Department, He worked on productivity and capacity enhancement project for mix model weld shop as a result of that, his team was able to reduce waiting time of Swift Desire in Market. He wrote 12 research papers in international and national peer reviewed journals, workshops and conferences. He has also set entrance paper for UPSC PCS Allahabad, India.

Email: mayank.agrawal@dei.ac.in

Professor C. P. Dewangan is currently working as a senior Associate Professor in the Department of Industrial and Production Engineering under School of Studies of Engineering & Technology at Guru Ghasidas (Central) Vishwavidyalaya (GGV), koni, Bilaspur, Chhattisgarh, India. He earned his B. E. Degree in Mechanical Engineering from Government Engineering College in the year 1992, Bilaspur, Chhattisgarh and earned his M.Tech in Industrial Engineering and Management from SGSITS, Indore (M. P) in the year 2001. He has an elevated Industrial experience of more than five years as a Maintenance Engineer in three reputed industries i.e. Khaitan Chemicals and Fertilizers Ltd. – Khargone (M.P), Comet steels Nanded -Maharashtra, Gajra Gears Ltd.- Devas (M.P). He started his academic carrier as a lecturer in the department of industrial and Production Engineering at Guru Ghasidas (Central) Vishwavidyalaya (GGV), Bilaspur in the year 2002. He has worked as a Head of Department (HOD) for more than ten years and acted as a chairperson for Board of Study of the department for more than six years. He has organized multiple national seminars, conferences, workshops under the dimensions of Industrial and manufacturing practices. He has reviewed three books under the shadow of repute publishers in the field of E.R.P, Metrology & Measurement and Operation Management. His area of interest includes Quality Control, Reliability, Work System Design, Supply Chain Management, Production & Operation Management, Operation Research, Industrial Engineering, Work Safety, Performance Measurement and so on.

Email: cpdewangan@yahoo.com

OUTCOME-BASED EDUCATION

For the implementation of an outcome-based education the first requirement is to develop an outcome-based curriculum and incorporate an outcome-based assessment in the education system. By going through outcome-based assessments evaluators will be able to evaluate whether the students have achieved the outlined standard, specific and measurable outcomes. With the proper incorporation of outcome-based education there will be a definite commitment to achieve a minimum standard for all learners without giving up at any level. At the end of the programme running with the aid of outcome-based education, a student will be able to arrive at the following outcomes:

PO1. **Basic and Discipline specific knowledge:** Apply knowledge of basic mathematics, science and engineering fundamentals and engineering specialization to solve the engineering problems.

PO2. **Problem analysis:** Identify and analyses well-defined engineering problems using codified standard methods.

PO3. **Design/development of solutions:** Design solutions for well-defined technical problems and assist with the design of systems components or processes to meet specified needs.

PO4. **Engineering Tools, Experimentation and Testing:** Apply modern engineering tools and appropriate technique to conduct standard tests and measurements.

PO5. **Engineering practices for society, sustainability and environment:** Apply appropriate technology in context of society, sustainability, environment and ethical practices.

PO6. **Project Management:** Use engineering management principles individually, as a team member or a leader to manage projects and effectively communicate about well-defined engineering activities.

PO7. **Life-long learning:** Ability to analyse individual needs and engage in updating in the context of technological changes.

COURSE OUTCOMES

After completion of the course the students will be able to:

CO-1: Have a good understanding of functioning of computer system

CO-2: Have a good understanding of the functioning of various subcomponents of computers.

CO-3: Student will be able to understand computing requirements for a specific purpose.

CO-4: Analyze performance bottlenecks of the computing device.

CO-5: Choose appropriate computing device for a given use case.

Mapping of Course Outcomes with Programme Outcomes to be done according tothe matrix given below:

Course Outcomes	**Expected Mapping with Programme Outcomes** *(1- Weak Correlation; 2- Medium correlation; 3- Strong Correlation)*						
	PO-1	**PO-2**	**PO-3**	**PO-4**	**PO-5**	**PO-6**	**PO-7**
CO-1	3	3	2	2	3	3	3
CO-2	3	3	3	3	2	3	3
CO-3	3	2	3	2	3	3	3
CO-4	3	3	2	3	3	3	3
CO-5	3	3	3	3	3	3	3

GUIDELINES FOR TEACHERS

To implement Outcome Based Education (OBE) knowledge level and skill set of the students should be enhanced. Teachers should take a major responsibility for the proper implementation of OBE. Some of the responsibilities (not limited to) for the teachers in OBE system may be as follows:

- Within reasonable constraint, they should manoeuvre time to the best advantage of all students.
- They should assess the students only upon certain defined criterion without considering any other potential ineligibility to discriminate them.
- They should try to grow the learning abilities of the students to a certain level before they leave the institute.
- They should try to ensure that all the students are equipped with the quality knowledge as well as competence after they finish their education.
- They should always encourage the students to develop their ultimate performance capabilities.
- They should facilitate and encourage group work and team work to consolidate newer approach.
- They should follow Bloom's taxonomy in every part of the assessment.

Bloom's Taxonomy

Level	Teacher should Check	Student should be able to	Possible Mode of Assessment
Create	Students ability to create	Design or Create	Mini project
Evaluate	Students ability to justify	Argue or Defend	Assignment
Analyse	Students ability to distinguish	Differentiate or Distinguish	Project/Lab Methodology
Apply	Students ability to use information	Operate or Demonstrate	Technical Presentation/ Demonstration
Understand	Students ability to explain the ideas	Explain or Classify	Presentation/Seminar
Remember	Students ability to recall (or remember)	Define or Recall	Quiz

GUIDELINES FOR STUDENTS

Students should take equal responsibility for implementing the OBE. Some of the responsibilities (not limited to) for the students in OBE system are as follows:

- Students should be well aware of each UO before the start of a unit in each and every course.
- Students should be well aware of each CO before the start of the course.
- Students should be well aware of each PO before the start of the programme.
- Students should think critically and reasonably with proper reflection and action.
- Learning of the students should be connected and integrated with practical and real life consequences.
- Students should be well aware of their competency at every level of OBE.

LIST OF FIGURES

CHAPTER 1

CHAPTER 2

CHAPTER 3

CHAPTER 4

CHAPTER 5

LIST OF TABLES

CONTENTS

1 PROCESS PLANNING AND PROCESS ENGINEERING

UNIT SPECIFICS

The following topics are covered in this unit:

- **Process Planning and Process Engineering:** Process Planning: Introduction, Function, Pre-requisites and steps in process planning, Factors affecting process planning, Make or buy decision, plant capacity and machine capacity. Process Engineering: Preliminary Part Print Analysis: Introduction, Establishing the General Characteristics of work piece, determining the principal Process, Functional surfaces of the work piece, Nature of the work to be Performed, Finishing and identifying operations.
- **Dimensional Analysis:** Introduction, types of dimensions, measuring the geometry of form, Baselines, Direction of specific dimensions. Tolerance Analysis: Causes of work piece variation, Terms used in work piece dimensions, Tolerance stacks. Work piece Control: Introduction, Equilibrium Theories, Concept of location, Geometric Control, Dimensional control, Mechanical control.

All the topics are discussed in a lucid manner along with their engineering applications. For student's practice, a moderate number of multiple-choice questions, short and long answer type questions and numerical problems are also given.

RATIONALE

To enhance quality, it is crucial to have a comprehensive understanding of the process. This chapter offers a comprehensive introduction to the concept of process and emphasizes the need of adopting a process-oriented mindset. Over the years, numerous approaches and philosophies have been developed to effectively manage and enhance processes. Comprehensive quality control. Lean and Six Sigma are particularly notable among these methodologies. Therefore, a thorough understanding of these principles is crucial for effectively coordinating process enhancement endeavors. The second part of this chapter explores the notion of Dimensional analysis, Tolerance analysis, and work piece control. These techniques are commonly used in engineering to simplify issue descriptions by lowering the number of variables and enabling more logical conclusions.

PRE-REQUISITES

- Fundamental Knowledge of any Manufacturing Industry Layout
- A course on Metrology and Measurements

UNIT OUTCOMES

After understanding this chapter readers will be able to

U1-O1: Understand the engineering and management concept of process planning.

U1-O2: Develop the concept of make or buy decision

U1-O3: Understand the concept of plant capacity and machine capacity

U1-O4: Understand the concept of print analysis and Dimension analysis

U1-O5: Understand the methods of workpiece control and dimension control

Unit-1 Outcomes	EXPECTED MAPPING WITH COURSE OUTCOMES (1- Weak Correlation; 2- Medium correlation; 3- Strong Correlation)				
	CO-1	CO-2	CO-3	CO-4	CO-5
U1-O1	3	2	3	3	1
U1-O2	1	2	3	2	3
U1-O3	3	3	2	2	2
U1-O4	2	2	1	2	1
U1-O5	2	1	2	2	3

1.1 Introduction to process:

"If you cannot describe what you are doing as a process, You do not know what you are doing"

W. Edward Deming

Deming's quote above elucidates the significance of process for companies. A company is a legal entity that is formed by individuals or groups of individuals to engage in business activities with the goal of making a profit. The definition is both straightforward and intricate. Corporate assets serve as a means of identifying a corporation in a straightforward manner. The corporate assets encompass a range of elements such as customers, workers, vendors, knowledge, physical assets, IT assets (software and hardware), monetary assets, operational intellectual property, strategic intellectual property, dynamic capabilities, culture, and business processes. The corporation possesses assets in both tangible and intangible forms. This is the initial level of complexity. The subsequent stage occurs when corporations must oversee these assets. In that situation, the corporation subsequently adopts many functional sectors with a shared goal of enhancing the financial outcome. Companies often encounter several processes that are necessary for managing different functional areas. Common management areas for a typical company include customer relationship management, human capital management, supply chain management, knowledge management, asset management, IT management, finance management, strategic management, operations management, project management, change management, and process management. Process and change management are the most formidable challenges that many companies encounter. When a company effectively implements change management and establishes a supportive culture, it becomes easier to facilitate process management. Once the organization is capable of visualizing the processes in various functional areas, the primary focus shifts to identifying or defining the processes. However, implementing this is more challenging than simply expressing it. Gartner's 2010 report highlights that one of the primary concerns for firms is the identification and enhancement of business processes. This demonstrates the intricacies that firms face when it comes to process identification. In order for firms to effectively prioritize process improvement activities, it is crucial for them to first define the processes involved. Process management is directly connected to both the process itself and, as a result, the overall success of a company. Therefore, firms must initially comprehend and establish processes.

1.2 Understanding Process

A series of related activities that circulate through an organization can be defined as a process. It may not be restricted to a specific department or function. The process must be presented in a manner that is readily observable from beginning to end.

A Process utilizes an organization's resources to deliver something valuable. It is essential to comprehend that the creation of any product and the provision of any service are impossible without a process. Conversely, every process requires a product or service to exist. Process Management involves the utilization of information, skills, tools, and procedures to effectively implement the Define, Measure, Analyze, Improve, and Control (DMAIC) methodology in order to fulfill client demands. Process management refers to the deliberate and systematic selection of inputs, processes, work flows, and methodologies that facilitate the transformation of inputs into desired outputs. For illustrative purposes, let's consider a bank as an example. A client is seeking approval for a loan. Therefore, the input will consist of a loan application. The process will entail a series of steps to authorize the desired output, which is the primary objective of the customer, namely the borrowed money. Additionally, there is an optional secondary output, which may include an email or a receipt confirming the approved loan. Therefore, we have acquired the knowledge that the supplier provides input in the form of information, while the consumer, who can be any entity that requests and consumes process outputs, receives the output.

The three basic vectors of a process management are: People, Process and Technology. An organization is unable to function without these basic vectors. The life cycle of process management encompasses the primary tasks required and highlights the ongoing process of enhancement. Therefore, this cycle of process management is continuously maintained and overseen over the whole lifespan of a company. Design includes the establishment of data parameters and the strategic planning of the process and functionality. Modeling processes involve the utilization of computer-generated software to simulate processes through multiple iterations, experimenting with various inputs and combinations until the most optimal performance is attained. An execution involves the implementation of processes and their integration with the current systems. Monitoring/control oversees the current status of business processes in real time and effectively manages the workload of tasks. It verifies for any potential mistakes. Optimization is required when additional processes are unable to yield significant enhancements in the process. The process is subsequently redesigned, involving the detection of bottlenecks and the enhancement of dashboards.

Process management has many benefits, such as better visibility, finding bottlenecks, optimizing, cutting down on lead time, and clearly defining roles. All of these things help any business or group run smoothly. There are some methods and tools that can be used to make a process better. Six Sigma methods and the Lean method are two of them. These methods try to make business and manufacturing processes better by first figuring out what the problem is, then making it clearer, finding the root cause, designing answers, checking how well they work, and finally making them all the same.

- A process is a set of actions that are necessary to accomplish the desired outcome of an operations system, as measured by a specific standard of effectiveness.
- When creating a product for a business, particular requirements are determined, including physical dimensions, tolerances, standards, and quality. Then it becomes a matter of determining the precise particulars of how to attain the intended outcome. This decision is the fundamental aspect of Process Planning.
- Process planning is the methodical process of determining the most efficient and competitive manufacturing procedures for a product.
- Process Planning is the subsystem that converts design data into work instructions.
- Process planning is the task at a manufacturing facility that identifies the procedures, process parameters, and equipment required to convert a piece-part from its original shape to a preset final shape specified on an engineering drawing. Process planning acts as an intermediate stage between product design and production (Fig. 1.1).
- The process planning begins when the product design is finished. However, in the initial phases of product design, it is essential to begin with the basic process planning. This includes the meticulous choice of materials and the determination of initial forms, such as casting, forging, and die casting. The drawing release is the ultimate manifestation of production design, providing a concise summary of the exact specifications for the planned construction. Process planning takes charge at this step and develops the overall production strategy for the part or product.
- Process planning utilizes drawings or specifications to determine the desired product and also considers projections, orders, or contracts to determine the quantity of products to be manufactured. The designs are subsequently examined to ascertain the comprehensive extent of the project. If the product is intricate and composed of multiple parts, a significant amount of labor may be required to disassemble it into its individual components and subassemblies.

- At this stage, first determinations can be made on the grouping of subassemblies to decide which parts should be manufactured and which should be purchased. Additionally, these judgments can help establish the overall level of investment required for tooling. Subsequently, a comprehensive routing plan is established for each individual component. Proficiency in understanding processes, machinery, and their capacities is essential, but equally important is a comprehension of production economics.

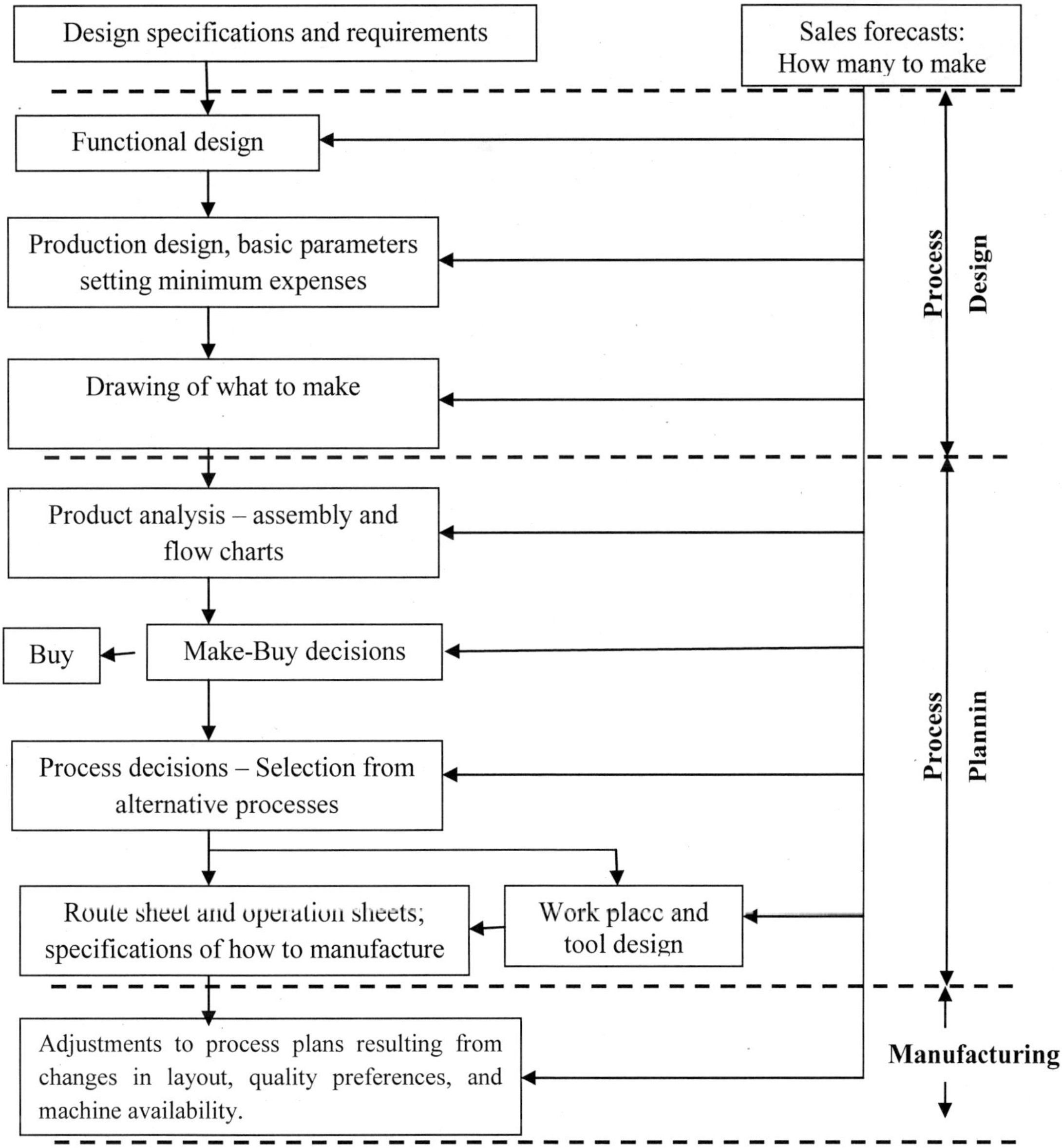

Figure 1.1 Comprehensive Advancement of Processing Plan

- Essentially, the engineering drawing of the component is analyzed in relation to the manufacturing process that will be employed. This stage is known as process planning and it involves the creation of a route sheet. The route sheet is a document that provides a detailed summary of the specific sequence of operations that need to be carried out on a particular component. The term "route sheet" is used because it includes a list of the machines that the part needs to go through in order to complete the series of operations.

1.3 Historical Background

Process management is a way to improve the performance of a business that focuses on the processes involved. It blends governance methods with information technology. Business process management makes the business processes of a company run more smoothly. As a result, it can be called a "process optimization process." Some people say that BPM makes organizations more flexible, effective, and able to change than a standard hierarchical management style that is focused on functions. The idea behind business process management is that people work together to do a set of tasks in a technology and organizational setting. Some business process tasks could be made better automatically with help from people using IT and information systems. The history of how business process management has grown:

1. Total Quality Management (TQM), a manufacturing methodology, was developed in the United States throughout the 1980s. Subsequently, Toyota, a renowned Japanese corporation, successfully implemented this strategy to manufacture superior products with the aim of capturing a larger share of the market. Consequently, TQM evolved into a sophisticated technique in the realm of process management (Jeston and Nelis, 2006).

Total Quality Management

Total Quality Management (TQM), a manufacturing methodology, was developed in the United States throughout the 1980s. Subsequently, Toyota, a renowned Japanese corporation, successfully implemented this strategy to manufacture superior products with the aim of capturing a larger share of the market. Consequently, TQM evolved into a sophisticated technique in the realm of process management.

Total quality management (TQM) is a comprehensive approach that aims to establish and maintain a culture within an organization where there is a constant focus on enhancing the organization's capacity to consistently provide customers with products and services of exceptional quality.

2. The approach of Six Sigma was developed in 1986 with the specific aim of emphasizing process change. Its objective is to enhance awareness regarding the significance of enhancing the process. (Jeston and Nelis, 2006).

Six Sigma

Six Sigma is a collection of methodologies and instruments used to enhance and refine processes. A six-sigma process is characterized by a statistical expectation that 99.99966% of all chances to generate certain characteristics of a product will be defect-free, with just 3.4 defective features per million opportunities.

3. Hammer and Champy (1990) introduced the concept of Business Process Reengineering (BPR) in the early 1990s. Business process re-engineering is a strategic approach to managing businesses that was first developed in the early 1990s. It involves analyzing and designing workflows and business processes within a company.
4. During the mid to late 1990s, Enterprise Resource Planning (ERP) systems gained widespread use. Several vendors have started offering enhanced techniques and "comprehensive problem-solving solutions" for organizations. Nevertheless, ERP systems may not adequately address the challenges faced by firms in managing their processes. They are unable to contribute to the enhancement of the process or optimize its efficiency and effectiveness, as suggested by Jeston and Nelis (2006).

Enterprise Resource Planning

Enterprise resource planning (ERP) refers to software used for managing company processes. It enables organizations to utilize a system of integrated applications to efficiently handle various back office operations relating to technology, services, and human resources. ERP software consolidates all aspects of a business, encompassing product planning, development, manufacturing, sales, and marketing.

5. During the late 1990s to the early 2000s, CRM systems that prioritized the customer's perspective and customer experience were initiated. These systems are unrelated to the back-office procedures (Jeston and Nelis, 2006).

Customer Relationship Management (CRM)

CRM is a strategic approach used by companies to effectively manage and oversee their connections with both existing and potential consumers. It frequently entails utilizing technology to arrange, streamline, and harmonize sales, marketing, customer service, and technical support.

6. In 2002, there was a significant surge in popularity for Business Process Management (BPM), commonly referred to as 'the third wave for BPM'. The work by Smith and Fingar (2002) generated considerable attention and sparked extensive discussion, making BPM the foremost and appealing subject in management.

Since 2003, Business Process Management (BPM) has regained popularity, leading to a rise in the use of Business Process Management Systems (BPMS) implementation plans. It has increasingly been implemented in our day-to-day business administration (Harmon, 2007). This favorable tendency aligns, to a certain extent, with the demands of marketing and commercial clients. However, the true underlying causes that have triggered this third wave of BPM implementation.

1.4 Concept of Lean

Lean means getting rid of steps in a process that aren't needed or add value all the time. The background of Lean goes back to the 1940s, when the Japanese company Toyota Automobile was about to go bankrupt. This is why Toyota is known as the "flag bearer" of Lean. Toyota went from being a small business to making the most cars in the world by studying the Indy 500, Ford Motor's Assembly Line (the idea of Flow: don't do steps that aren't needed), and grocery stores (the idea of Pull). It has successfully combined the ideas of Flow and Pull and is based on the strict principle that waste kills.

Lean was built on the idea of standard work, which doesn't always mean similar work. It means that people should look at their work and figure out how to do it in a way that meets the needs of all shareholders the best. Lean also cares about level loading, which makes work go more smoothly. Lean can be used to define value, find the stream, make it run, let the customer pull, and strive for perfection. Flow, or the constant movement of production units, is at the heart of Lean. It is thought that there will always be steps in a process that don't add value. Because of this, turnaround time for each process has gone down. Lean is based on four main ideas: a) getting rid of waste (which happens naturally in any manufacturing process; some examples are transport, inventory, motion, waiting, over-processing, over-production, defects, and skill); b) adding value to the process and getting rid of activities that don't add value; c) listening to the customer and meeting their needs; and d) optimizing the process, with the idea that perfection can never be reached.

The Kaizen method is part of lean methods. It is based on the idea of SCORE, which stands for Select - Clarify - Organize - Run - Evaluate. Lean is mostly used for problems that are short or not too complicated, and it is based on time. Lean is all about getting rid of steps and practices that aren't needed. Lean makes sure that we're doing the right thing.

1.5 Concept of Six Sigma

Six Sigma is a set of tools and methods for making things better. To make manufacturing and business processes better, mistakes must be found and fixed, and attention must be paid to outputs that are important to customers. Six Sigma, in simple terms, makes sure that we do the right thing the first time we do it. Six Sigma's goal is to make money by making customers happy. This can be done by cutting down on mistakes, difference, and inconsistent service. Motorola was the first company to use this method, in 1986.

Six Sigma means different things in different situations. As a statistical tool, Six Sigma methods only let 3.4 mistakes happen out of every million chances. While the DMAIC method is not a process itself, it is a technique that is used to get better results. It's a way of thinking that says anything that isn't perfect can be made better and flaws cost money. Greater the sigma number, the better the process works and the fewer mistakes it makes. The main goals of a Six Sigma project are to cut down on differences, stop mistakes by putting the process at the center, and meet customer needs. There won't be a known solution to a good Six Sigma project's main cause or complex problem. The project will also take three to six months to finish, and there will be clear defects and problems throughout the process.

Sigma capacity	%Yield	Defects per million
2	69.15	308537

Sigma capacity	%Yield	Defects per million
3	93.32	66807
4	99.38	6010
5	99.98	233
6	99.99966	3.4

Table 1.1 Relation between Sigma capacity and Defects

The core principle of Six Sigma is to optimize customer satisfaction by minimizing the number of defects per million chances (DPMO), maintaining process focus on the target, and reducing variances in performance (standard deviation). The role of management in a Six Sigma Project is to comprehend the issues, provide budgetary commitments, choose initiatives, establish a review process, pick champions, and identify strategic strategies. The DMAIC method is utilized for the current process.

1.6 DMAIC

DMAIC is an acronym that represents the five stages of a problem-solving methodology: Define, Measure, Analyze, Improve, and Control. It is a methodology employed to get a superior degree of performance. Stands is an acronym that stands for the definition of a problem and the research of the customer's needs. The text provides a quantifiable description of the problem and outlines the underlying method for determining how performance will be measured. The process of measurement involves the use of metrics or measures to analyze performance and identify areas for improvement through statistical methods, such as the Process Capability Ratio and Index. Analyze refers to the process of identifying the genuine root cause(s) of the underlying problem. The system gives priority to potential causes and subsequently suggests solutions. Enhance the process of identifying and testing the most effective innovations that target the underlying issues. Put simply, it chooses the most optimal approach for improvement that is both simple and economical. Control refers to the process of identifying sustainable solutions that guarantee the maintenance of enhanced process performance. The process entails the selection of a matrix, monitoring the progress, implementing a full-scale solution after testing, finalizing the solution, and establishing new norms.

1.7 Process Capability

Process Capability refers to the capacity of a process to achieve the design specifications, which are defined as the desired targets for the service or goods. It assists managers in achieving and sustaining a consistent process distribution in terms of both mean and variation. The control limits are determined by the mean and variation, rather than the design specification. The primary goal of a statistical process control system is to generate statistical signals when there are underlying causes of changes. The process capability index quantifies the level of effectiveness of a process. Put simply, it is a metric that evaluates how well a process performs in relation to the desired outcome.

Process Capability Analysis is utilized to forecast defect rates, facilitate quantity enhancement, and assess the efficacy of disparate processes. The measurement of Capability should only be conducted upon the completion of the process. Process Capability can be utilized in both administrative and production-oriented systems. Graphically, a process is considered competent if it falls inside the range defined by the lower specification limit and the upper specification limit.

1.7.1 Process Capability Ratio (Cp)

The ratio between the distance of the lower specification and the upper specification, in relation to Six Sigma, must be equal to or larger than 1.0 for a capable process. In a Six Sigma process, the ratio is precisely 2.0. However, the text does not provide any indication of the process's position, specifically whether it falls within the specified range. The basic principle states that the majority of values in any process distribution are contained within the range of -3 sigma to +3 sigma. If the ratio is below 1.0, specifically $Cp < 1.0$, then the product or services are considered to be beyond the acceptable tolerance level. However, the processes are not always positioned optimally within the specification range. Therefore, one additional parameter is necessary.

1.7.2 Process Capability Index (C_{pk})

The Process Capability Index is a measure of the proportion of the process's inherent tolerance (3 sigma) between the process mean and the nearest specification limit. Put simply, it pertains to the concept of being centered. A process is considered to be fully competent when its Cp value is greater than 1.0 and its C_{pk} value is also greater than 1.0. The goal of the index is to maintain the average near the center. A process is considered centered when the Process Capability Index for both the Upper and Lower Specification Limit is equal to unity. If the Cpk value is greater than 1, it indicates that the distribution of data is becoming more concentrated and the goal value is increasing. However, if the C_{pk} value is less than 1, it means that the process does not match the specified requirements.

1.8 Factors affecting Process Management

1.8.1 Business Process

The process that any business or business department follows to fulfill their customer's needs or complete their assigned tasks.

1.8.2 Management Tools

The instruments utilized by company managers to strategize, implement, and assess the efficacy of their business operations.

a. **Business/Strategic direction:** The business's strategic objectives and future-oriented limitations. This tool is commonly presented as annual budgets, capital plans, business objectives, and similar forms.

b. **Forecast:** Projected forecast of the company's anticipated level of business. Forecast time periods span from a weekly to a yearly basis, and exhibit variations in terms of specificity and organization.

Forecasts can predict the expected dollar income, the quantity of units, or the number of operational activities that will be needed depending on the sales of the company's products.

c. **Performance plan**: The translation of performance standards that a firm must achieve in order to meet the forecasted goals while adhering to the limits of the business direction. For instance, if a company predicts that it will sell 100 widgets in a month and the budget sets a target of Rs. 5000 for operational expenses, the performance plan for the month should aim to reach a cost per unit of Rs. 50.

d. **Schedule:** A specific and detailed strategy outlining the individuals, tasks, and timeline that the organization will follow in order to accomplish the objectives outlined in the Performance Plan.

e. **Operations control:** This is the point where proactive planning and retrospective evaluation intersect. It is the instrument that enables a manager to promptly determine whether they are now progressing towards meeting the established timetable, which was designed to accomplish the performance objectives derived from the forecast and business strategy.

f. **Results measurement:** measuring success in a way that looks backwards. These tools help managers find differences and make operations better by showing them the parts of the process that aren't meeting the Schedule... Plan for performance, a forecast, and the direction of the business.

g. **Continuous improvement loop:** A group of tools that keep track of changes in performance and connect operational tasks to results in the financial world.

1.8.3 Management Behavior and Mindset

The utilization of management tools by business managers to direct the business process.

a) **Silo thinking** refers to a mindset found in certain corporations where specific departments or sectors are reluctant to share information with others within the same organization.

b) **Rigidity** : It is ill-advised to permanently establish the framework, as it should be delicate.

c) **A lack of recognition** can demotivate individuals, whereas receiving appreciation for their efforts encourages them to sustain and enhance their performance with increased passion.

d) **Intuition**: It is imperative to exclude emotions from business decisions and rely on trustworthy analytical data.

1.9 Key Issues of Process Management

Process management encompasses a range of tools and technologies, including business analyst tools, workflow engines, business activity monitoring software, business rule engines, user interface systems, and document or control interaction systems. Each of them is utilized to carry out duties and operations. However, due to the diverse range of organizational styles and intricate settings, challenges and issues arise.

a. Implementing a process management system entails a significant financial investment, and it can be challenging to accurately determine its potential to enhance the company's competitiveness and generate profits in the near term.

b. The complexity of process management makes it challenging to determine the potential success of utilizing BPM.

c. What are the essential components for initiating, constructing, and maintaining a prosperous process management program?

d. How can process management enhance organizational agility in response to both internal and external change?

1.10 Information required to do Process Planning

- Amount of work to be completed, as well as the special requirements of the product.
- Level of workmanship required.
- Availability of equipment, tools, and manpower.
- Order in which procedures will be executed on the raw material.
- Equipment names for performing the operations.
- The prescribed duration for each individual task. At what time will the operations be conducted?

1.11 Process Planning Procedure

These are the different steps that go into planning a process:

1. Preparation of working drawings
2. Deciding to make or buy
3. Selection of manufacturing process
4. Optimizing machine capacity and selecting the appropriate machines/equipment
5. Material selection and bill of materials
6. Choice of jigs, fittings, and other attachments
7. Operational planning and tooling prerequisites
8. Creation of papers such as operation and route sheets, among others.

1.12 Working Drawing

Working drawings are integral documents in manufacturing, serving as comprehensive guides for producing components or products. These drawings meticulously detail the geometric shape, dimensions, tolerances, machining requirements, surface finishes, heat treatments, coatings, bill of materials, and any other pertinent information necessary for manufacturing and inspection processes. Assembly drawings present the entirety of the product, illustrating all parts in their correct relationships, while subassembly drawings focus on specific groups of parts within the product. Detail drawings provide in-depth information for each individual part, including dimensions, material specifications, and other crucial details. In manufacturing companies, much of the planning revolves around these mechanical drawings, serving as the foundation for various departments. The tooling department, for instance, relies on these drawings to strategize the necessary tooling for the manufacturing process.

Standardization is crucial in creating drawings to ensure efficiency and clarity in manufacturing processes. Each drawing should adhere to a standardized format, facilitating quick identification of part numbers and ensuring that every item depicted in the drawing is assigned its unique number and name, corresponding to the parts list. Additionally, comprehensive details regarding material standards must be incorporated for all materials, arranged in a standardized manner to avoid overlooking important specifications and to streamline the review process. Important notes such as testing requirements should not be relegated to obscure locations on the drawing but rather prominently displayed in a standardized panel on the drawing template, ensuring essential information is readily accessible and minimizing time spent on deciphering complex drawings.

Before the process engineer looks over any plan, they should make sure of the following:

- Do dimensioning and datum surfaces align with commonly used machining techniques?
- Are adequate stock allowances supplied on castings, forgings, and stampings to account for expected mismatch or distortion during heat treatment?
- Is there enough clearance and access provided to ensure the proper assembly of all components?
- Are the tolerances on functional qualities practical and is statistical tolerancing employed whenever feasible?
- Are sufficient clamping and locating surfaces supplied for manufacturing?

1.13 Make or Buy Decision

Decisions on whether to produce or purchase materials, parts, or assemblies should be guided by well-informed recommendations. The provided information should be detailed enough to facilitate intelligent decision-making processes.

1.14 Factors affecting Make or Buy decision

1.14.1 Quantitative Factors

(1) Opportunity costs: Opportunity cost can be defined as the financial worth that is given up while choosing one option over another. The facilities used in manufacturing a part or component are essentially forfeited for any other purpose. The choice between making or buying often comes down to an effort to maximize the utilization of resources.

(2) Incremental costs: Typically, only the expenses that change based on the choice to produce or purchase are considered relevant.

(3) Idle facilities: The presence of unused facilities has a direct impact on the decision to produce internally or outsource, especially when it comes to calculating the additional expenses. When there are enough resources, just the costs that change based on the amount produced, known as variable costs, need to be taken into account.

1.14.2 Qualitative factors

(1) Product quality: One may choose to manufacture parts in their own factory to ensure better control over the overall quality of the final product, notwithstanding the potential cost savings of purchasing the parts externally.

(2) Patents: Legal constraints may impose limitations on a company's ability to manufacture specific components.

(3) Skills and materials: Some parts may require highly technical expertise or rare and specialized materials, making it impossible to manufacture them in-house.

(4) Long-term Considerations: In the short term, it might be more profitable to use idle resources to make more parts in-house, for example during slow times. But this could lead to bad relationships with suppliers or even run out of parts from outside sources during busy times when the company's own buildings could be used in more profitable ways.

1.14.3 Other factors

The intangible factors that may impact the decision to make or buy include:

- Total count of external vendors.
- Dependability of external sources.
- Periodic requirements

Example 1.1: M/s Benso Pvt. Ltd. is engaged in the production and sale of gas stoves. It manufactures certain components for the gas stoves and procures the remaining parts. The engineering department is considering the feasibility of producing one of the parts that is now being bought for Rs. 8.50 per, in order to reduce costs. Annually, the company utilizes 100,000 units of these components, and the accounting department assembles the subsequent cost breakdown based on engineering approximations.

- The fixed costs will go up by Rs. 50,000.
- The cost of labor will go up by 125,000 rupees.
- The factory's costs, which are currently Rs. 500,000 a year, could go up by 12%.
- The cost of the raw chemicals needed to make the part is Rs. 600,000.

Based on these figures, should M/s Benso Pvt. Ltd Company make the part or keep buying it?

Answer: Calculate total cost incurred if the part was manufactured:

- Additional fixed costs = Rs. 50,000
- Additional labour costs = Rs. 125,000
- Raw materials cost = Rs. 600,000
- Additional overhead costs = Rs. 60,000 (0.12 x 500,000)
- Total cost to manufacture = Rs. 835,000
- Cost to manufacture one part = Rs. 835,000/100,000

  = Rs. 8.35

Hence, it is advisable for the Company to produce the part internally as the manufacturing cost per part (Rs. 8.35) is lower than the current purchasing cost (Rs. 8.50 with a difference of Rs. 0.15 per part).

1.15 Process Selection

Process selection determines how the product (or service) will be produced. It involves

1. Major technological choice
2. Minor technological choice
3. Specific component choice
4. Process flow choice

1.15.1 Major technological choice

When considering the development or acquisition of a product, several critical questions arise regarding technology

- Is there existing technology to manufacture the product?
- Are there any alternative technologies from which we need to select?
- Should the technology be domestically developed?
- Should other countries be granted licenses for innovations?

1.15.2 Minor technological choice

Once the main technological choice has been made, there may be a few smaller technological process options that can be used instead. The operations manager should help figure out which options are the most cost-effective and will still meet the goals for product and capacity.

- To minimize costly starts and shutdowns, the steel and chemical industries should implement a continuous process that operates 24 hours a day.
- Conversely, an assembly line process adheres to the same sequence of processes as mass production, but it is not required to operate continuously for 24 hours a day. Examples of sectors that utilize this approach include car and ready-made clothing manufacturing.
- Job shop methods manufacture things in small batches, potentially tailored to specific customers or markets.
- Let's consider selecting a work store. There are further options available. For instance, within an industrial setting, the process of fabricating, connecting, and finishing two metal pieces may constitute a negligible fraction of the overall production of a final product. There are multiple methods for casting and molding, as well as various techniques for cutting, shaping, assembling, and finishing.
- Choosing the optimal combination of procedures based on cost and overall operational efficiency can be challenging.

1.15.3 Specific component choice

When determining the type of equipment and level of automation to employ in a given task or industry, several factors must be considered. Firstly, the nature of the task itself plays a crucial role; some tasks may require specialized equipment for precision or efficiency, while others may benefit from more versatile, general-purpose machinery. Additionally, the degree of automation should be tailored to strike a balance between productivity and the preservation of human roles. While automation can enhance efficiency and reduce labor costs, it's essential to assess its impact on employment and ensure that human oversight is maintained where necessary for safety and quality control. Ultimately, the decision regarding equipment specificity and automation levels should be guided by a comprehensive analysis of the task requirements, economic considerations, and ethical implications.

Computer-aided manufacturing (CAM) and industrial robots are becoming more prevalent in many manufacturing systems.

1.15.4 Process flow choice

Optimizing the flow of products inside an operations system is a complex and essential activity that aims to maximize efficiency and minimize costs. The ultimate process-selection step plays a crucial role in coordinating the smooth flow of materials and products through the system. Assembly drawings, charts, route sheets, and flow process charts are essential tools for a full analysis of the process flow. By using these technologies, procedures can be carefully analyzed and adjusted to optimize material handling and save storage costs. This may involve reordering, combining, or even eliminating certain processes. This strategic strategy improves both operational efficiency and cost-effectiveness, resulting in optimized overall system performance.

The four phases of process selection, as previously mentioned, are intricately interconnected. During each phase, decisions should be taken to minimize the costs associated with process activities.

1.16 Factors affecting process selection

A process is essential for the purpose of molding, shaping, conditioning, and assembling materials and components using machinery and labor to transform raw materials into a final product. One should choose the most cost-effective method (a breakeven chart, similar to the one presented in Chapter 3, may be helpful for this purpose) and sequence that meets the product parameters.

The choice of process is contingent upon:

1.16.1 Current production commitments

If a sufficient amount of work has already been assigned to more efficient equipment, it may be necessary to transfer the current workload to less efficient machines in order to accomplish it within the designated time frame.

1.16.2 Delivery date

An early delivery date may:

- Force the use of less efficient machines,
- Rule out the use of special tools and jigs as they will take time for design and fabrication.

1.16.3 Quantity to be produced:

The high expense of preparation and efficient setups will likely not be justified by the little quantity. As a result, it's conceivable that less efficient machines will have to make them, and vice versa.

1.16.4 Quality standards

Quality requirements can restrict the option of producing the product on a specific machine, among other factors.

1.17 Plant Capacity

Capacity refers to the pace at which production is produced, the amount of output generated within a specific timeframe, and it represents the maximum achievable quantity of output within that timeframe. However, capacity is a concept that is both dynamic and can be altered and controlled. It can be somewhat adapted to accommodate changes in sales levels.

Determining plant capacity is a critical aspect of operations management that involves assessing the maximum output a facility can sustainably achieve within a given timeframe. This capacity is influenced by various factors such as equipment capabilities, workforce efficiency, production processes, and market demand. Understanding plant capacity requires a comprehensive analysis of these factors to ensure optimal utilization of resources while meeting customer needs. Capacity planning involves forecasting demand, evaluating existing resources, and identifying potential bottlenecks or constraints (explained in chapter 2). By strategically managing plant capacity, businesses can enhance productivity, minimize costs, and maintain competitiveness in the market. Regular reviews and adjustments to capacity planning are essential to adapt to changing market conditions and ensure long-term success.

1.18 Machine Capacity

Machine capacity refers to the maximum output that a specific machine or piece of equipment can produce within a given time frame under optimal conditions.

- Machine capacity can be described as the amount of time that is available for work on a machine, measured in machine hours or minutes. As an illustration, a machine could have a maximum weekly capacity of 168 machine hours, which is equivalent to 7 days of 24 hours each.
- To determine the scheduled machine capacity, the additional hours worked in overtime should be included in the normal machine capacity on a weekly basis.
- Currently, it is important to deduct time for maintenance (period when the machine is not operational), determining the duration by consulting the maintenance plan and a statistical log of previous machine failures.
- Furthermore, it is necessary to allocate sufficient time for machine downtime (when the machine is not in use due to waiting for work or lack of an operator) as well as for machine auxiliary tasks (such as setting up and cleaning).
- In order to determine the actual amount of time available for productive work, adjustments must be made to the machine operating time. These adjustments account for any deviations in average performance from the standard performance used to establish the standard timings for work processes.

- The ultimate capacity attained after implementing all these adjustments is referred to as the Standard Machine running time, which is typically far lower than the maximum of 168 hours per week.
- Table 1.2 depicts the different degrees of machine capacity. It can be determined that it is feasible to assess the load against the capacity at various levels by making adjustments to compensate for time lost, either to the load or the capacity.

<table>
<tr><td colspan="6">Maximum machine capacity: 168/week</td></tr>
<tr><td colspan="3">Normal machine capacity</td><td colspan="2">Planned overtime</td><td>Not worked</td></tr>
<tr><td colspan="5">Planned machine capacity</td><td></td></tr>
<tr><td>Machine running time (planned)</td><td colspan="2">Ideal machine time (forecast)</td><td>Machine ancillary time (forecast)</td><td>Machine downtime (forecast)</td><td></td></tr>
<tr><td colspan="4">Machine running time (planned)</td><td colspan="2"></td></tr>
<tr><td colspan="2">Standard machine running time</td><td colspan="2">Low performance (forecast)</td><td colspan="2"></td></tr>
</table>

Table 1.2 Types of capacity

1.18.1 Analysis of Machine Capacity

The process of obtaining accurate information regarding the capacity of the available machines to produce the desired output is known as machine analysis.

- An objective of machine analysis is to obtain the answers to certain definite questions in regard to the use of manufacturing machines.
- What is the duration required for a specific machine to complete its operation on a single unit of material?
- What is the daily, weekly, or monthly processing capacity of this equipment in terms of material units?
- What is the daily maximum capacity for each process on each material at the plant?
  The initial query can be resolved in either of two ways.
- Through the utilization of standard data, one can acquire information either through direct experimentation and testing, or by consulting records of previous achievements.
- The second question can be resolved by determining the machining time and set-up time, and by accounting for the necessary idle time.
- To answer the third question, the overall plant capacity in units of product is determined by aggregating the number of units that can be processed by machines with similar capabilities.

- Based on this data, it is feasible to ascertain the utmost capacity of each individual process and the entire facility. Machine load charts, which display the tasks assigned to each machine, can also be generated.

Several ratios pertaining to this subject matter include:

1. ***Machine availability*** = $\frac{\text{Machine available time}}{\text{Total machine time}} \times 100$
2. ***Machine utilization*** = $\frac{\text{Actual running time}}{\text{Machine availble time}} \times 100$
3. ***Machine efficiency*** = $\frac{\text{Standarf running time}}{\text{Actual running time}} \times 100$
4. ***Machine effective utilization*** = $\frac{\text{Standarf running time}}{\text{Machine available time}} \times 100$

1.19 Process and Equipment Selection Procedure

The process and equipment selection technique entails formulating a comprehensive description of manufacturing activities and constructing a preliminary process for each specific feature defined by the product designer.

Before starting the selection of the preliminary procedure, more inputs are needed.

➢ In order to effectively plan and manage costs, it is necessary to set targets for facility and piece costs. This involves specifying the raw materials required, determining the hourly production volume before determining machine capacity, establishing the schedule, and selecting the appropriate process.

As part of the preliminary process selection, the production engineer will calculate the number of steps and subsequent stations required to incorporate all the design features specified on the blueprint. To do this, it will be necessary to visually represent each individual sequence, determine the amount of personnel needed, and make an approximate estimation of the required layout provisions to support each phase of the process.

1. After finishing the initial processing steps, the manufacturing engineer should create a list of alternative processes, especially for areas that have been identified through detailed analysis of the initial processing as having high costs, uncertain performance, or where the likelihood of meeting the requirements of the individual operator is considered to be low. The past data pertaining to similar activities can be helpful in generating these opportunities.
2. An exhaustive and systematic comparison of each stage of the provisional process with each stage of the alternative process will enable the production engineer to choose the most balanced position that maximizes cost, quality, flexibility, and inherent risk.

Assuming all engineering management and manufacturing factors are the same, production procedures will be selected based on the most advantageous return on investment or other financial measures.

3. Once the process selection is finished, it is conveyed to the product engineering, industrial engineering, plant and maintenance engineering, industrial relations, and finance departments.

Effective coordination and communication are crucial for the successful integration of new technologies into the current plant and staff.

4. Engaging in meticulous processing. Once the procedure has been chosen and shared with all relevant departments, the final intricate processing, which is crucial for all subsequent activities, is began.

During the intricate processing, the production engineer will heavily rely on the vast amount of engineering knowledge possessed by the machine and equipment suppliers.

5. The initial step in developing the detailed processing is to gather the most up-to-date information regarding product design, manufacturing rate, facility cost, and part cost targets, similar to the establishment of the provisional process. The detailed processing procedure is similar to the provisional processing, with the exception that each step of the process will be thoroughly identified and documented using process estimate sheets.
6. Each phase, from rough stock to final product, includes information about the origins of the material. Each process sheet should clearly indicate the subsequent operation to ensure a systematic flow of parts through the production activities.

1.20 Selection of Material, Equipment, etc

Material selection is crucial due to the vast variety of materials and their forms, ensuring the material's quality, chemical composition, and size limit scrap generation while achieving the desired product shape.

1.20.1 Bill of Material

Bill of material is also known as parts list. Bills of material or specification sheets are the most common ways to break down a product into its parts.

A bill of materials tells you what you need to buy and what you need to make. It needs to say whether the part is to be bought or made. The production-control department utilizes the bill of material to ascertain manufacturing and scheduling dates. Process engineering use it as a comprehensive list to ensure the completion of their tasks.

Methods engineering utilizes it in the calculation of time allowances for assembly procedures. Niel The shops department creates accumulations based on the bills of material. They subsequently established the lists of shortages for the expediters in the production-control department to utilize. The finished-stores section releases assembly units based on the bills of material.

The structure of the bill of material may differ slightly in minor aspects, depending on how individual companies utilize it. The typical information requested on the bill of materials form comprises

- The Product/Model Name
- Model code for identification
- Application
- Date of preparation

- Name/initials of preparer
- Name/initials of checker
- Part numbers
- Make/purchase designations
- Subassembly/Child part numbers and names
- Quantity requirements
- Material used in each part.

1.20.2 Selection of Jigs, Fixtures and Other Special Attachments

The presence of these supporting devices is essential for two reasons: to increase the production rate and to decrease the cost of production per piece.

1.20.3 Selection of Cutting Tools and Inspection Gauges

They are essential for the following purposes:

- Minimize manufacturing time;
- Conduct precise and efficient inspections.

1.20.4 Create a process layout

That clearly outlines each operation and the specific sequence in which they are to be executed.

1.21 Process Analysis

Process Analysis means the study of the overall process in a factory (plant). It analyses each step of the manufacturing process and aims at improving the industrial operations. Process analysis aids in finding better methods of doing a job and this is achieved by eliminating unproductive and unnecessary elements of the process or through modified layout of facilities.

The process is analyzed with the help of Process Charts and Flow Diagrams.

The process analysis involves several distinct steps:

- Choose the procedure for examination.
- Divide the process into individual operations and further break them down into sub-operations.
- Create a process chart and a flow diagram.
- Examine the process chart and flow diagram by putting each stage to a questioning method.
- Revise the process chart and flow diagram to accurately represent the amended (proposed) procedure.
- Evaluate the suggested approach to verify all the asserted benefits.
- Provide a comprehensive explanation of the novel approach to the workforce and implement it.

1.22 Operation Planning and Tooling Requirements

Operation planning is a critical phase in the planning process, signifying the conclusion of the process planning stage. The process entails meticulously strategizing the specifics of the approach to carry out each task at the designated workstation and devising the required equipment. Operation sheets serve the purpose of displaying the intended order of work items and conveying crucial information for the production of each part. They have comprehensive plans for every component, subassembly, and final assembly, outlining the path that parts will take through different departments, the specific order of operations, the necessary machinery, specialized tools, and gauges, the time needed for each operation, and specific specifications regarding speeds and feeds.

The process planner is responsible for designing and determining the necessary tools for each operation. These tooling requirements are documented to ensure that the relevant tools can be easily located when needed. The record may have a basic inventory of tools for the tool store, or, in cases of high production volume, include specialized operation sheets for both tool stores and machine setters.

An optimal approach to labeling tool kits for each operation is to incorporate the part number together with a suffix, or to employ a classification system that clearly indicates the relationship between the tooling kit and the production process. After the operation sheet is approved and completed, the actual execution of plans can commence, enabling the assembly and inventorying of the necessary machines, tools, gauges, fixtures, and jigs for fabrication.

1.22.1 Manual Process Planning

This form of planning is referred to as man-variant process planning and is the most prevalent method utilized for production in modern times.

In order to plan the operations for producing a part, it is necessary to have knowledge of two sets of variables: the specifications of the part (as represented by an engineering drawing) and the equipment and processes that are available, together with the capabilities of each process.

Based on these variables, the planner chooses the optimal combination of operations needed to manufacture a final product. When choosing this mix of processes, several criteria are used.

The primary factors considered in process selection are often production cost and time. However, the utilization of machines and the routing of tasks also frequently impact the chosen plans. Typically, the process planner aims to choose the optimal combination of processes and machines to manufacture an entire range of parts, rather than simply one individual item.

1.22.2 Automated Process Planning

Man-variant process planning can sometimes become a monotonous and laborious task. It generates inaccurate process planning. The labor-intensive nature of human-driven planning, along with this factor, has prompted numerous sectors to explore the possibility of automating process planning.

An entirely automated process planning system would remove all human intervention from the moment an engineering drawing is produced until the generation of a detailed process plan for every manufacturing operation.

Advantages of computer-aided process planning

- It has the capability to decrease the amount of time required for process planning.
- It can decrease the level of expertise needed by a planner.

- It has the ability to generate more reliable and uniform blueprints.
- It has the capability to generate more precise plans.
- It has the potential to enhance productivity.
- It has the ability to decrease both the expenses associated with process planning and manufacturing.

Various sectors have recorded the advantages of computer-aided process planning. These systems have the capability to significantly decrease the amount of time required for planning, either from days to hours or from hours to minutes.

1.23 Print Analysis

In the manufacturing industry, print analysis usually refers to the inspection and assessment of technical drawings, blueprints, or specifications concerning the manufacturing of a specific part, product, or assembly. Examining intricate plans or diagrams that list the measurements, tolerances, components, finishes, and other requirements needed to manufacture a part are part of it.

Print analysis is used to make sure that everyone involved in the manufacturing process understands the design intent and that the finished product satisfies the necessary specifications and quality standards. Early on in the production process, this analysis can help identify possible problems or challenges, enabling adjustments to be made to reduce waste, increase efficiency, and prevent errors.

The following are Print Analysis's goals:

- Give the specific details from the part print.
- Determine the workpiece's functional surfaces.
- Choose the order in which the workpiece will undergo the various operations.

1.24 Functional surfaces of the work piece

The surfaces that need to be created on the workpiece during the manufacturing process are referred to as functional surfaces. Imagine a workpiece that needs to be machined. It is necessary for the machined surfaces to mate with other machined surfaces. There are three main methods to identify the areas that need to be machined from the part prints.

- Surface finish
- Basic geometry
- Tolerances

Studying the part print shows what needs doing. It tells us how to do each step to make the part just right. The part's features also decide how easy or hard and cheap or expensive it is to make. So, the process engineer checks these features in the print to plan well. They figure out how the part's traits affect the work. Then, they match these traits with the tasks ahead. This way, they make smart choices to use resources well and do the job efficiently.

In this context, responding to the following questions can be helpful:

- To what extent is the workpiece symmetrical?
- What is the number of relationships between machined surfaces?

- What connection exists between these surfaces?
- Is it possible to group or combine related surfaces or areas in order to minimize the number of machining setups?
- How many steps need to be taken on each surface?

1.25 Dimension Analysis

In engineering science, measurements are often made by comparing them to an agreed standard, such as the "meter" or "millimeter" scale. This method uniquely describes the physical quantity, but it can also result in different values depending on the applied unit. To quantify a physical quantity, it is necessary to choose a comparative measure. One solution is to compare the regarded length to another relevant length scale. This method provides a precise numerical value and becomes independent of whether the unit is "millimeters" or "meters." This approach is applicable to all physical quantities and their respective units.

This approach leads to meaningful statements and conclusions, as it reduces the number of variables in the problem description. For example, if the problem is governed by two lengths, relating them to one another reduces the number of variables, simplifying the problem description and allowing for easier solution discovery. Additionally, fewer experiments are required to reveal the complete physical conduct. Dimensional analysis encompasses the precise formulation of this simplified approach, as well as its scientific reasoning and generalization.

The method of dimensional analysis is generally valid. It always provides accurate results. The method is applicable in all branches of physics, whether in mechanics, fluid mechanics, thermodynamics, electricity, optics or other areas. It is useful in chemistry and biology, it is of great value in engineering and is indispensable in model theory and experimentation. In many areas, the systematic application of dimensional analysis has led to great progress, such as in problems of heat transfer and fluid flow engineering. Dimensional considerations were particularly fruitful and fundamentally improved insight in the field of turbulent flows. Certainly, in the application of dimensional analysis one should not proceed carelessly! The simplicity of the method sometimes leads to believe that its sole use alone already brings physical insight. But that is certainly not the case. It is rather necessary to deeply investigate the problem, to abstract and simplify. Only when we clearly understand the dominant physical mechanisms for a specific problem and insignificant effects are left aside, dimensional analysis provides valuable support, as it allows reducing the number of variables.

From a historical point of view, dimensional analytical approaches date back to the origins of mechanics. Great progress was made in 1686 by Newton and his fundamental work on the motion of bodies (see Görtler 1975). The classical mechanics, founded by him, is based on the three fundamental quantities length, mass and time, which were regarded as fundamental and independent of each other. The notion of physical dimension was only introduced in 1822 by Fourier. It took another 50 years until in 1873 von Helmholtz determined the essential dimensionless products for hydrodynamics in a systematic manner. Ten years later, Reynolds published his studies on the transition of laminar to turbulent flow and related this to the Reynolds number which perhaps represents today's best known dimensionless number (see Görtler 1975). Towards the end of the 19th and beginning of the 20th century the research activity in the field of dimensional analysis tremendously increased. It probably did not enter into general consciousness until 1914 when Buckingham established its general formulation and justification in the essay "On physically similar systems", Buckingham (1914). The master of dimensional analysis is undoubtedly Lord Rayleigh. In his essay "The principles of similitude" published in 1915 he gives numerous examples of the application of dimensional analysis, Rayleigh (1915). Amongst others, on the basis of dimensional considerations he explains why the sky appears to be blue. Since the time after Buckingham's publication, dimensional analysis became a firm pillar in the building of physics. It has been included in textbooks and curricula of natural and engineering sciences. It has as already mentioned-made significant contributions to the development of many fields of knowledge and is of particular importance in engineering applications.

1.25.1 Physical Quantities

Physics and engineering sciences deal with laws that connect different physical quantities. In general terms physical quantities denote some measurable characteristics or properties of physical objects (objects, states or processes). The objects themselves are not necessarily measurable, but their characteristics are. So, for instance, a room per se cannot be measured, yet its height, width, length or volume can be measured. Just as well, a mass point is not measurable by itself, but its coordinates, its velocity or its mass can be quantified. Further, heat conduction is not measurable, but a temperature distribution, the temperature gradient or the thermal conductivity represent measurable quantities. The term measurable quantity also includes variables that cannot be measured directly, but are determined from other measured quantities (Table 1.3).

Physical quantity	Symbol	Dimensions
Velocity of sound	as	LT-1
Acceleration	a	LT-2
Angular velocity	ω	T-1
Diffusion coefficient	D	L2T-1
Density	P	ML-3
Dynamic viscosity	η	ML-1L-1
Energy	E	ML2L2
Frequency	f	T-1
Force	F	MLT-2
Heat transfer coefficient	H	ML-3θ-1

Physical quantity	Symbol	Dimensions
Power	P	ML2T-3
Pressure	P	ML-1T-2
Resistance force	W	MLT-2
Surface tension	σ	MT-2
Thermal conductivity	k	ML2 T-3ϑ-1
Thermal diffusivity	ά	L2T-1
Velocity	U	LT-1

Table 1.3 Some often-used physical quantities and their dimensions

1.25.2 Entities, Base Quantities, Derived Quantities and Quantity System

The qualitative property of a physical quantity is called its entity. For example, the duration between the maximum deflections of a pendulum is of the entity time. Also, the duration in which a cyclist travels a certain distance is of the entity time. The height of a space, the diameter of a sphere, as well as the distance between two points all belong to the same physical entity, which is the length. Physical entities that are independent of each other are called base quantities. The entirety of all base quantities that are necessary to describe the regarded physical relations are referred to as quantity system. Generally, in physics the base quantities for length L, mass M, time T, temperature, electric current I, amount of substance N and the luminous intensity J are used as basic set of quantities. The choice of this [LMTOINJ] system of quantities is by no means mandatory, but rather represents pure convention. The appropriate choice of a quantity system according to the regarded problem is discussed later. Based on these base quantities the so-called derived quantities may be deduced, either resulting from physical laws or resulting from useful definitions. Hence, the derived quantities are power law products of the base quantities. As an example, consider the quantity of density, which is the ratio of mass to volume, thus mass divided by the cube of the length. Another example is quantity for velocity, which follows from length divided by time.

1.25.3 Units, Unit System and Unit of Measurement

While the qualitative property of a physical quantity is described by its entity, the quantitative value is determined by a numerical value (measured value) and a unit. Numerical value and unit are thereby correspondingly multiplied. The units of the base quantities are called base units. Today, the [LMTOINJ] system of quantities is generally applied in terms of the corresponding SI-system "Système International “'Unités". This system was established in 1960 internationally and consists of the base units meter (m), kilogram (kg), second (s), Kelvin (K), ampere (A), mole (mol), and candela (cd) (see DIN 1305 from 1974). The unit of a physical quantity is indicated by curly braces around the symbols. for example,

$$\{u\} = \frac{m}{s} = m^1 s^{-1}$$

Entity	Symbol	SI-unit	Symbol
Length	L	Meter	M
Mass	M	Kilogram	Kg
Time	T	Second	S
Temperature	θ	Kelvin	K
Current	I	Ampere	A
Substance amount	N	Mole	Mol
Luminous intensity	J	Candida	cd

Table 1.4 Base qualities and base unit in the SI-Unit-System

The entirety of the base units is called unit system, and the entirety of the base quantities and base units is referred to as measuring system. In Table 1.4 the base quantities and base units of the SI-system are summarized.

1.26 Industrial Inspection

Industrial inspection is examining components manufactured during production to determine if their dimensions conform to the allowed limitations specified by the consumer. Inspection is categorized into two types: active and passive. Active inspection involves the examination of parts during their manufacture in a manufacturing system. Online inspection, also known as real-time monitoring, allows for immediate corrective action in the event of any issues. However, in passive inspection, parts are examined throughout the production process and the purpose of inspection is to separate the components into categories of good or bad. In most industries, there is a broad encouragement for active inspection, although its implementation may vary depending on the specific application.

Components that are manufactured by any production method adhere to a normal distribution. The machine environment can be conceptualized as an endless universe, whereas the machined components are regarded as a collection of random data. When data is collected in a random manner from an infinitely large universe, the resulting data conforms to a normal distribution, as depicted in Figure 1.2. Let's look at an example to help us understand what it means. It is possible to make shafts with a width of 25.0 mm on a lathe machine. There will be differences in the sizes of the rods as they come out of the machine, such as 25.002 mm of length, 24.985 mm of width, and so on. The reason for this is that these measurements have a normal distribution. People say that no two parts made from the same production method will be the same, so you can't make a copy of the part. As shown above, X is the machine's mean or target number, and is the process dispersion, also known as the standard deviation.

Here are some things that could change the process or the desired mean (X) number in a normal machine setting:

1. Tool and die deterioration.
2. Excessive oscillations of the machine.
3. Wear or minor failure or loosening of machine parts.
4. Machine and process modification.

Measurement inaccuracy or alteration in material characteristics.

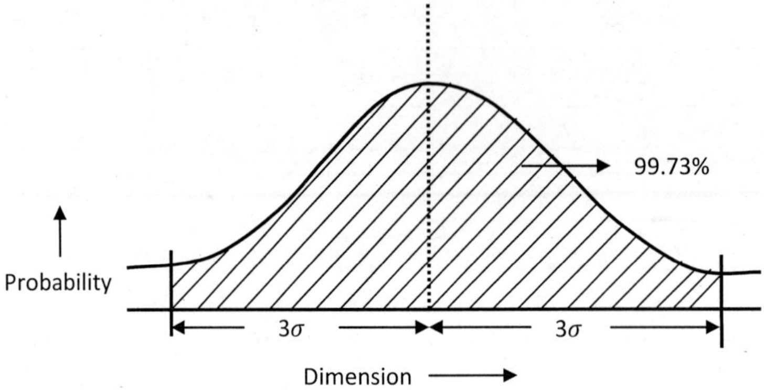

Figure 1.2 Normal Distribution Curve

The standard deviation of the process is influenced by the following elements.

1. Operator negligence.
2. New or inexperienced operator.
3. Machine's condition has worsened.
4. Regular resetting or recalibration of the procedure.
5. Enhance the diversity of the material.

Approximately 99.73% of the area under the normal distribution curve lies within the limits of ± 3 standard deviations (σ) from the mean. This implies that 99.73% of the components will be manufactured within the 6 σ range. The term "process capability" refers to the ability of a machine to consistently produce output within the 6 σ standard deviation. In order to proceed with the inspection, let us gain a clear understanding of the concept of normal distribution through the use of the following instances.

Example 1.2: The lathe machine is designed to manufacture shafts with a diameter of 25.0 mm. This conventional lathe has a process dispersion of 0.15 mm. Find out:

1. What is the production quantity of shafts below 24.9 mm by the machine?
2. What is the production count of shafts with a diameter more than 25.2 mm?
3. What is the typical range of shafts produced by the machine?
4. Given that the allowable acceptability range for shafts, as determined by the consumer, is 25 ± 0.35 mm, the number of defective shafts produced by the machine may be determined.

Solution: Problems pertaining to normal distribution can be resolved by utilizing normal distribution tables provided at the appendix of the book. The tables are presented with respect to a variable Z.

$$Z = \frac{X - \bar{X}}{\sigma}$$

X is the value up to which the total area under the normal distribution from - ∞ to X is to be calculated as shown in the given Fig. 1.3.

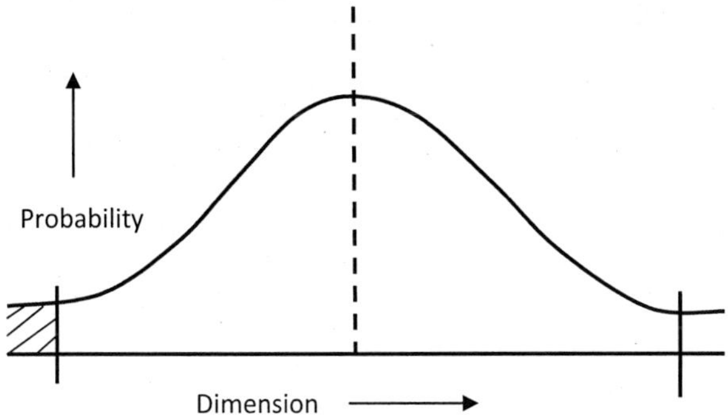

Figure 1.3 Standard Normal Distribution Curve

1. In the given problem we have to calculate $= \int_{-\infty}^{24.9} P(X)dX$

Where P(X) is the probability function

$$X = 24.9$$

$$\bar{X} = 25$$

$$\sigma = 0.15$$

So,

$$Z = \frac{X - \bar{X}}{\sigma} = \frac{24.9 - 25}{0.15} = \frac{-0.1}{0.15} = -0.666$$

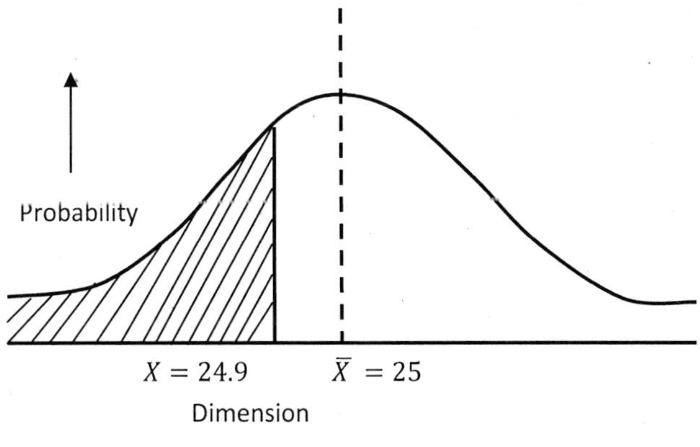

Figure 1.4 Normal Distribution Curve for x=24.9

2. The area under the normal distribution up to a Z value of 24.9 mm is 0.2546. Approximately 25.46% of the shafts will have a diameter below 24.9 mm, while approximately 74.54% of the shafts will have a diameter more than 24.9 mm.
3. In the above problem the shaded area has to be calculated as shown in Fig. 1.6. The area corresponding to this value from the table is 0.9082. So, the shaded area is (1-0.9082) =0.0918. So, 9.18% will be dimension more than 25.2 mm.

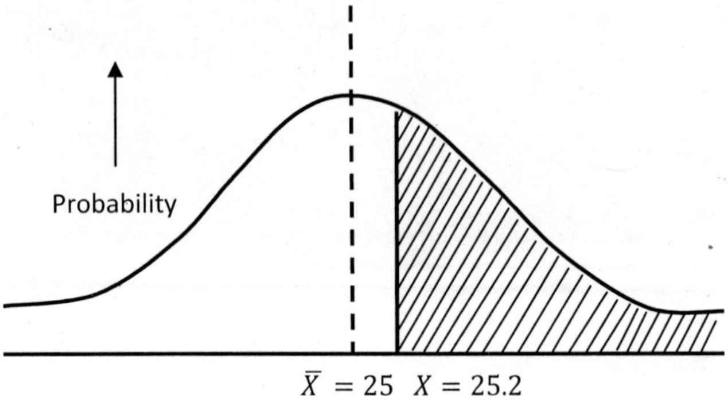

Figure 1.5 Normal Distribution Curve for x = 25.2

4. Approximately 99.73% of the components will be manufactured with a tolerance of 25 + 3(0.15) mm. The machine will manufacture components within the range of 24.55 to 25.45 mm.
5. The shaded region represents the faulty components that were manufactured by the machine. Given the symmetrical distribution, the calculation of defective items on one side can be obtained by multiplying the computed area by 2. The area under the given distribution is less than 24.65 mm$Z = \frac{24.65-25}{0.15} = 2.33$

Area from the table corresponding to Z = -2.33 is 0.0099. So, shafts below 24.65 are 0.99%

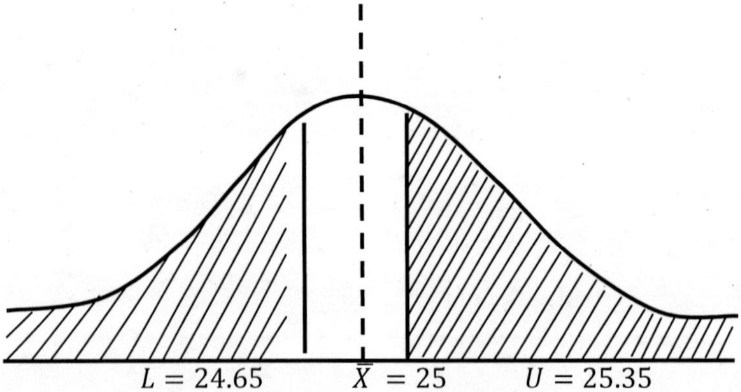

Figure 1.6 Normal Distribution Curve

Total defective percentage = 2 × 0.99 = 1.98%

Example 1.3: A machine at a party can sprinkle fragrant water on each guest's head as a way to thank them. One chance observation shows that the machine sprinkles 4 grams of fragrant water. 0.04 grams is the normal range for the machine. The machine injects 4 grams of water every 2% of the time. What is the average amount of water it injects?

Solution:

$$Z = \frac{4 - \bar{X}}{0.04}$$

Th Value of Z corresponding to 0.02 are= -2.055 (From Normal distribution Table)

$$-2.055 = \frac{4 - \bar{X}}{0.04}$$

$$\bar{X} = 4.0822$$

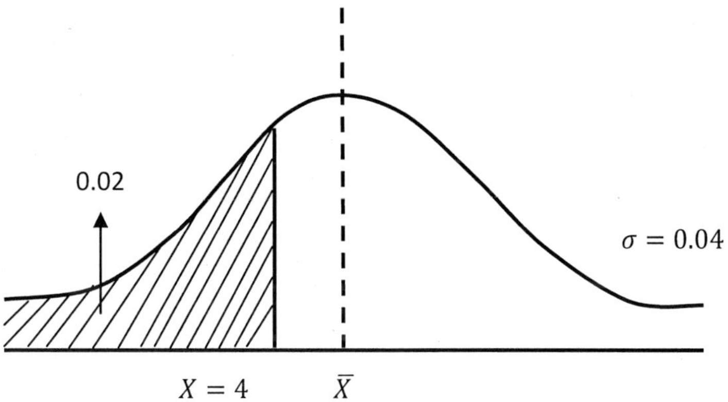

Figure 1.7 Normal Distribution Curve

1.27 Accuracy and Precision

Accuracy means being close to the exact number, and precision means being able to do it again and again. To be precise, you need to know the process dispersion, while accuracy is about the average value of the things you make. The average value of the observations in Fig. 1.8 (i) is very close to the desired value, but there are big changes in the observations because the process is more powerful. This type of system is accurate but not exact. The standard deviation of the observations in Fig. 1.8(ii) is low, and the mean is close to the desired number. This means that the system is precise and accurate. In Figure 1.8(iii), the process spread of the system is low, but the mean value is very far from the desired value because the machine wasn't set up correctly. This kind of method is called precise, but it's not really accurate. A lot of differences exist in Fig. 1.8 (iv), and the mean value is very far from the desired value. People say that these kinds of methods are neither precise nor accurate.

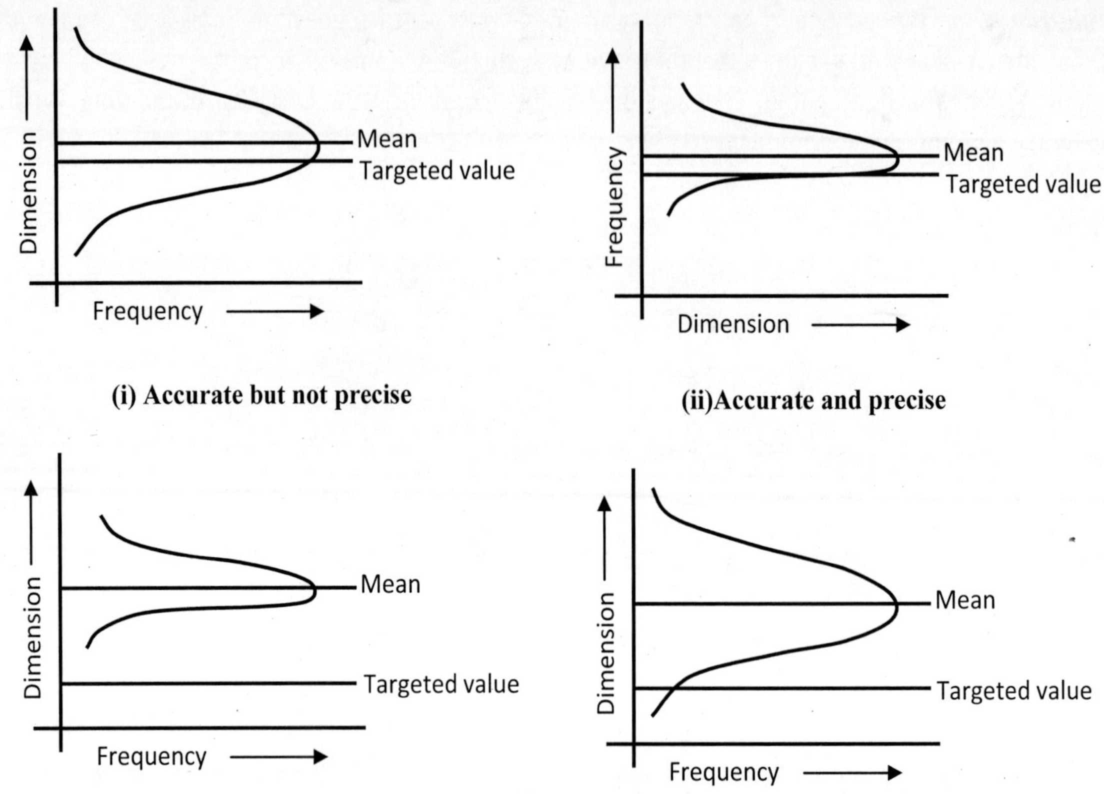

(i) Accurate but not precise **(ii)Accurate and precise**

(iii) Precise but not accurate **(iv) Neither precise nor accurate**

Figure 1.8 Accuracy Vs Precision

1.28 Interchangeability

An assembly consists of a combination of a hole and a shaft. A hole refers to any internal characteristic of a part, which may or may not be circular. Similarly, any external characteristic of a part is referred to as a shaft, which may or may not be circular. Typically, the hole is created first due to the availability of standard-sized drills and reamers. If the shaft is created before the holes, the holes are referred to as "made to suit". As previously said, duplicating the part is not feasible, hence there are restrictions on the allowed dimensions. The discrepancy between the upper and lower limits is referred to as the intended tolerance by the consumer. The designated value of the hole and shaft is referred to as the fundamental size, whereas the closest whole number according to the standard is known as the nominal size.

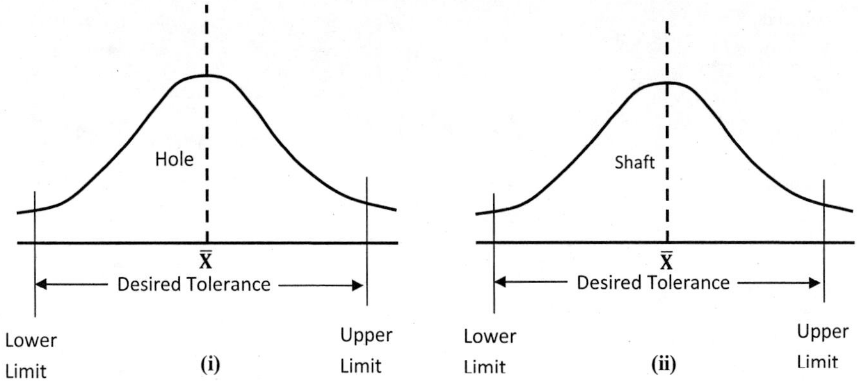

Figure 1.9 Interchangeability

Assume that there are two distinct machines manufacturing holes and shafts. If the process capacity is equal to the desired tolerance for each individual component (as depicted in Figure 1.9 (i) and (ii)), it is possible to assemble a randomly selected hole from the hole lot with a randomly selected shaft from the shaft lot. These systems are referred to as totally interchangeable systems that do not necessitate any examination following machining. Interchangeability offers several advantages:

1. The cost of assembly falls due to the ability to produce holes and shafts at locations with lower costs for materials and labor.
2. It is feasible to standardize the dimensions of a hole and a shaft.
3. The product's quality improves.
4. The maintenance of the assembly is cost-effective and convenient since it is not necessary to discard the entire assembly when either the hole or shaft fails. Only the failing parts need to be replaced.

1.29 Limits, Fits and Tolerances

As discussed earlier that due to process capabilities it is not possible to duplicate any part, so for assembly parts (either hole or shaft) should come within certain range of dimensions. This acceptable variation is called tolerance. The term "upper limit" refers to a larger size, while the term "lower limit" refers to a smaller size.

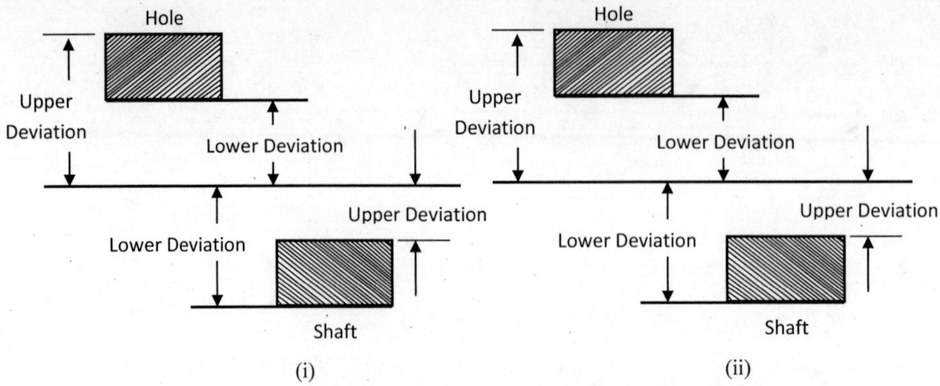

Figure 1.10 Tolerance and Fundamental Deviation

The fundamental deviation refers to the distance from the basic size where the tolerance zone is located. The upper limit is called E_{cart} Superiors (ES) (a French name) and the lower limit is called E_{cart} Inferieur (El). Fig. 1.11 illustrates a total of 25 distinct forms of fundamental deviation.

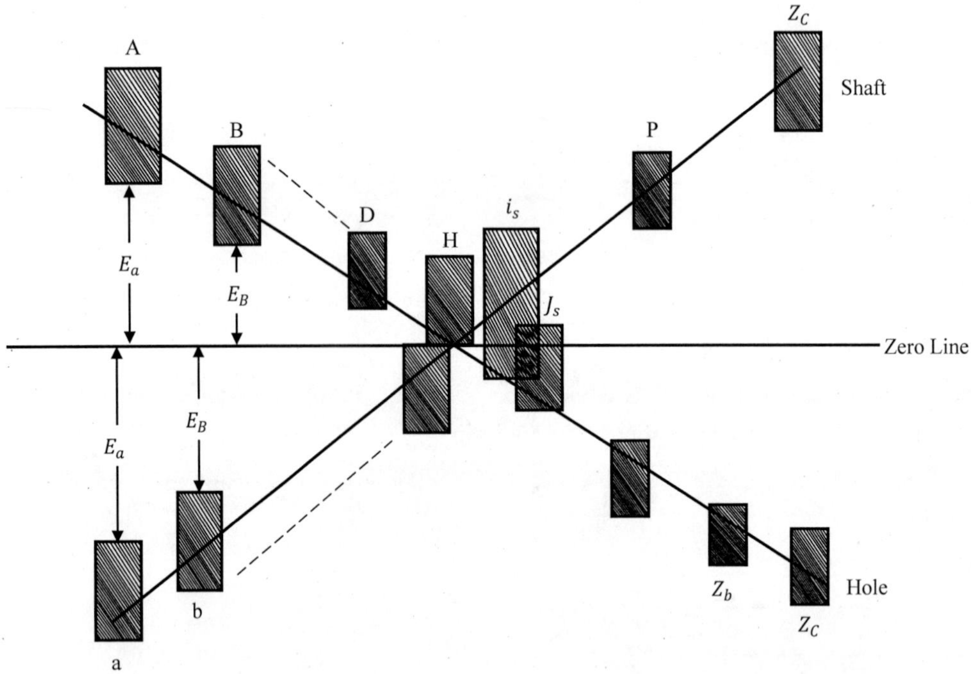

Figure 1.11 Variation of Different Fundamental Deviation Types

There are omitted letters in the alphabet and additional letters, such as js, Za, Zo, and others. The fundamental deviation value for a hole of 'A' type will be equivalent to that of a shaft of 'a' type, but with opposite polarity. All basic deviations can be represented using empirical formulas. An assembly is formed by the conjunction of a hole and a shaft. If the assembly contains a hole of type 'H', it is referred to as a hole basis system. Conversely, if the assembly contains a shaft of type 'h', it is referred to as a shaft basis system.

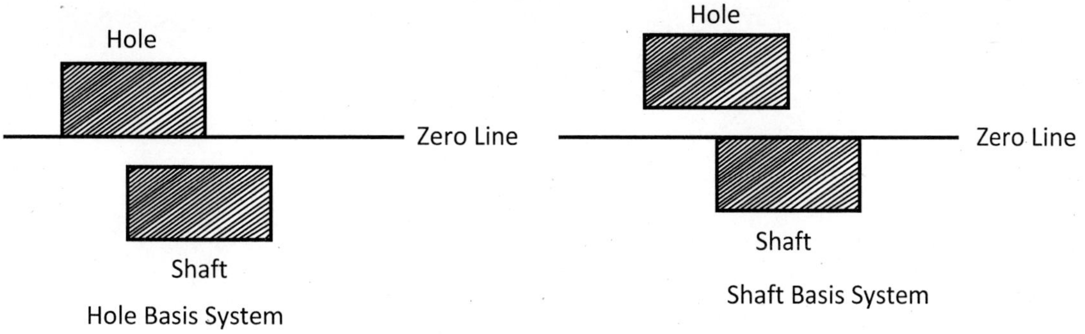

Figure 1.12 Hole and Shaft basis system

Fundamental deviation is also expressed as the distance between zero line and the limit closer to it i.e. either El or eg. Information presented in capital characters denotes the presence of a hole, while information provided in lowercase letters denotes the presence of a shaft.

It is generally understood that a larger shaft and a smaller hole will result in a greater amount of material. The restrictions that define the maximum size of a shaft or the minimum size of a hole are referred to as the Maximum Material Limit and the Minimum Material Limit, respectively.

A fit is the correlation between the dimensions of a hole and a shaft prior to their assembly.

Broadly there are three types of fits:

1. Clearance fit: If the lower limit of the hole is greater than the higher limit of the shaft, as depicted **in Figure 1.13.** It is classified as a clearance fit.

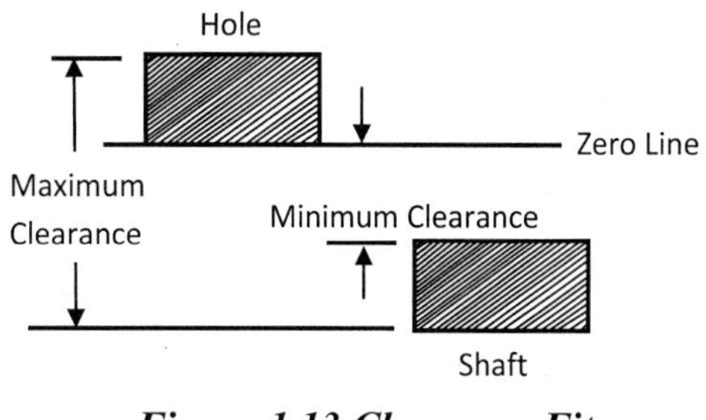

Figure 1.13 Clearance Fit

2. Transition fit: This form of fit occurs when there is an overlap in the tolerance zones. Physically, this shows that when a part is randomly picked from a group of parts with holes and a group of parts with shafts, certain assemblies can be produced without applying force, while others will require force.

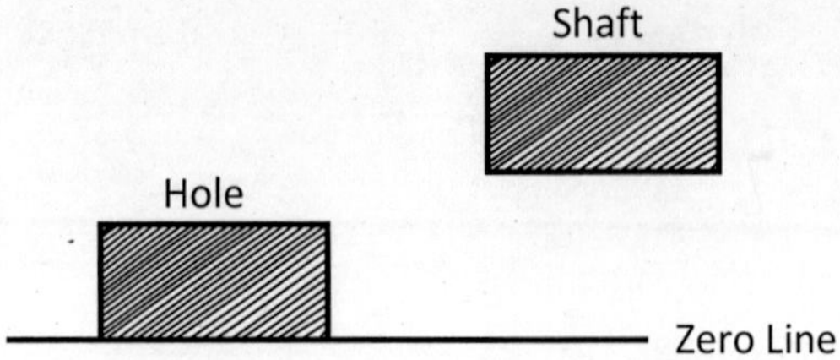

Figure 1.14 Transition Fit

3. Interference fit: If the maximum diameter of the hole is smaller than the minimum diameter of the shaft, force must be exerted in order to assemble them. These types of fittings are referred to as interference fits.

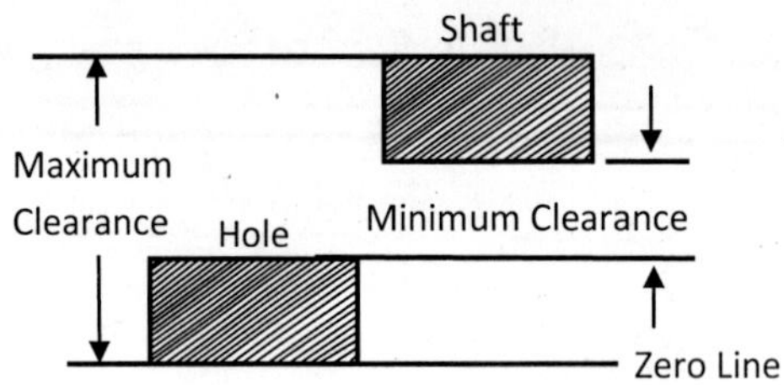

Figure 1.15 Interference Fit

1.30 Allowance

It is defined as the difference between maximum material limit of hole and shaft. Depending upon the type of fit either it is equal to minimum clearance or maximum interference.

1.30.1 Width of Tolerance Zone

There are three systems used in the world.

1. British Standard BS-4500-1969
2. International Standard ISO: 286-1988
3. Indian Standard IS-919

However, the criteria for fundamental deviation and the range of acceptable tolerance are same. As previously mentioned, there are 25 standardized basic variances. The initial recommendation for the tolerance width was to adjust the tolerance based on a cubical expression.

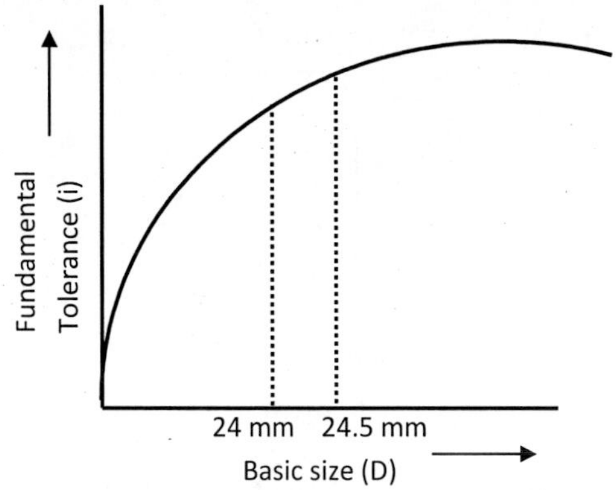

Figure 1.16 Fundamental Tolerance

$$i = 0.45\sqrt[3]{D} + 0.001D$$

From Figure 1.16, it is evident that when the fundamental size rises, the tolerances will correspondingly increase. It is not feasible to precisely manage the tolerance for a bigger basic size. However, a significant issue arises when the fundamental dimension changes by a small amount, such as from 24 to 24.5 mm, resulting in a modification of the tolerance values. Subsequently, it was determined that there would be uniform tolerances within a specific range of diameters. The diameters are standardized.

Above (mm)	Up to and including (mm)	Above (mm)	Up to and including (mm)
-	3	80	120
3	6	120	180
6	10	180	250
10	18	250	315
18	30	315	400
30	50	400	500
50	80		

Table 1.5 Diameter Range for Same Tolerance

The curve representing the fundamental tolerance 'i' intersects with the geometric mean of the diameter range, as depicted in Figure 1.16. The geometric mean diameter will be utilized in all tolerance and basic deviation formulas. Originally, there existed 16 levels of tolerance zones. Two additional tolerance ratings were introduced for precision industries. These tolerances were arranged in the five step preferred series i.e. every fifth step will be multiple of 10 and the series is approximate geometric series. The calculation is shown in the Table 1.5.

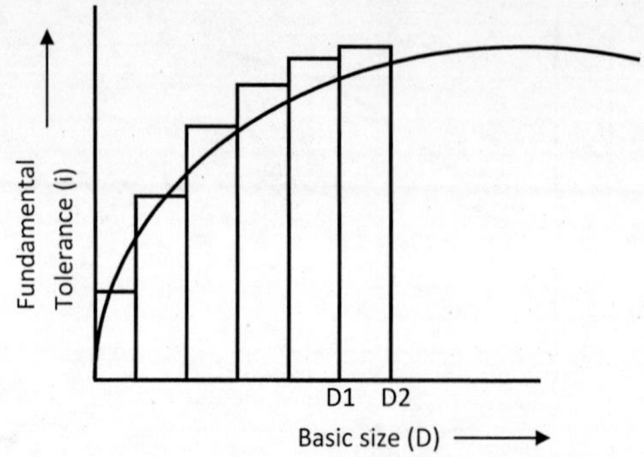

Figure 1.17 Modified Fundamental Tolerance

The name "modified 5 step preferred series" comes from the fact that this series doesn't follow any rules. At present

- IT 01 to IT 1 => Empirical formula
- IT 1 to IT 5 => Exact geometric series
- IT 5 to IT 16 => Preferred series

Suppose a hole and shaft assembly is designated by $25H_7d_8$. 25 indicates the basic size 'H' is the type of hole with width of tolerance zone IT 7. 'd' is the type of shaft with IT 8 is the width of tolerance zone. This can be represented by following fit diagram.

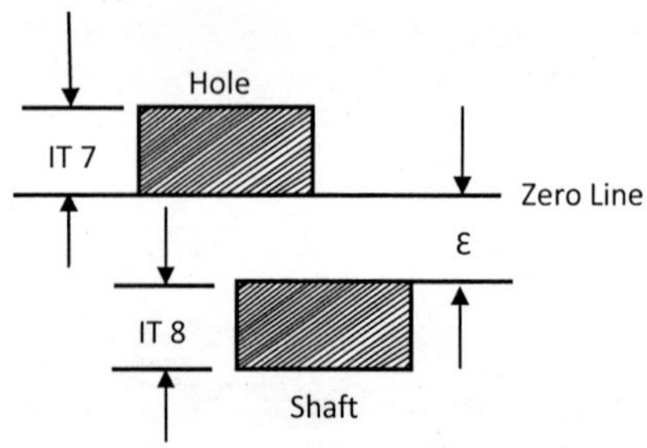

Figure 1.18 Fit Representation of 25H7d8

T 01 0.3+0.008 D	IT 0 0.5+0.012 D	IT 1 0.8+0.02D i a	IT 2 ar	IT Ar^2
IT 4 Ar^3	IT 5 $(10)^{4/5}i=6.31\ i$ $Ar^4 \Rightarrow 7i$	IT 6 10 i	IT 7 $10(10)^{1/5}i=$ $15.8\,i \Rightarrow 16\,i$	IT 8 $10(10)^{2/5}i=$ $25.11\,i \Rightarrow 25\,i$
IT 9 $10(10)^{3/5}i= 40i$	IT 10 $10(10)^{4/5}i= 60i$	IT 11 100 i	IT 12 160 i	IT 13 250 i

IT 14 400 i	IT 15 640 i	IT 16 1000 i		

Table 1.6 Tolerance Series

When the tolerance zone is only on one side of zero line, it is called unilateral. When tolerance zone is on both side of zero line it is called bilateral.

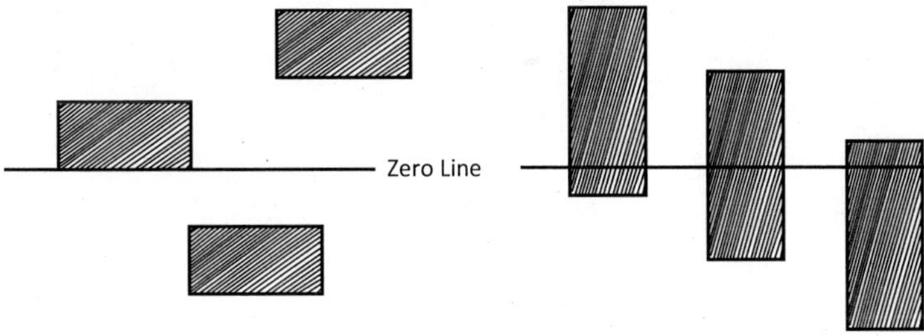

Figure 1.19 Unilateral and Bilateral Tolerance

Example 1.4: A hole and shaft have a basic size of 30 mm and are to have a clearance fit with maximum clearance of 0.04 mm and a minimum clearance of 0.02 mm. The hole tolerance is to be 1.5 times the shaft clearance. Find out limits for both hole and shaft using:

- A hole basis system
- A shaft basis system.

Solution:

1. Hole basis system

$X + 0.02 + 1.5\,X = 0.04$

$X = 0.008$ mm

$1.5x = 0.012$

Hole = Max = 30.012 mm

Min = 30.012mm

Shaft Max = 29.98 mm

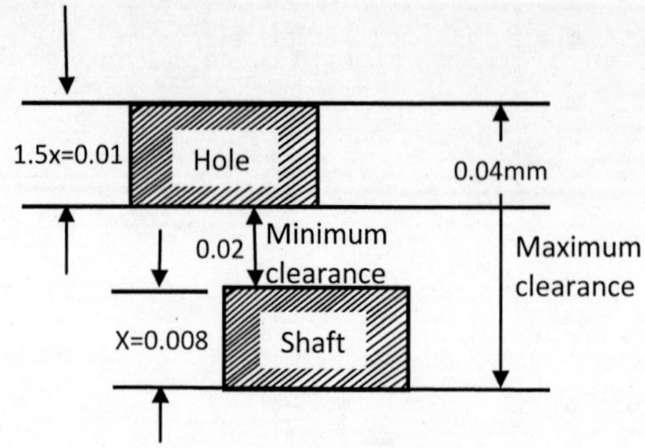

Figure 1.20 Hole Basis

2. **Shaft basis system**

X + 0.02 + 1.5 X = 0.04

X = 0.008 mm

1.5 X = 0.012 mm

Hole →Max = 30.032 mm

Min = 30.02 mm

Shaft → Max = 30.00 mm

Min = 29.992 mm

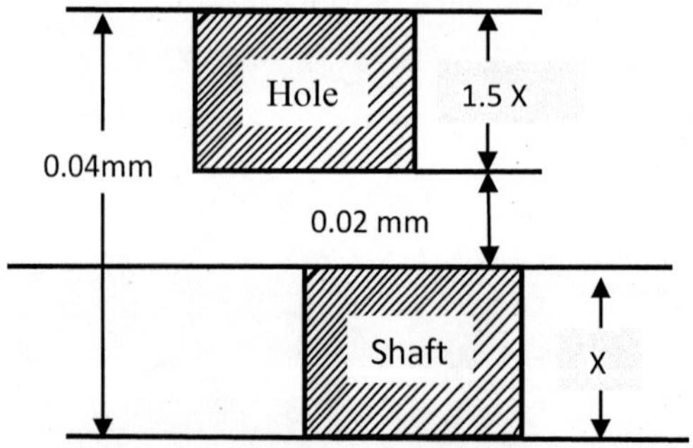

Figure 1.21 Shaft Basis

Example 1.5: Calculate the fundamental deviation and tolerance and hence the limits of the size for the shaft and hole for the following fit 64 mm H_8f_7. The diameter steps are 50 mm and 80 mm. For shaft description f, upper deviation is supposed as -5.5 $D^{0.41}$:

Date: For	Tolerance
H8	25i
F7	16 i

Solution: Since 64 falls between diameter range of 50 to 80 so,

$$D = \sqrt{50 \times 80}$$

= 63.24 mm

Fundamental tolerance i $= 0.45\ D^{1/3} + 0.001\ D$

= 1.856 microns

We should note here that although input D is in mm but is in microns.

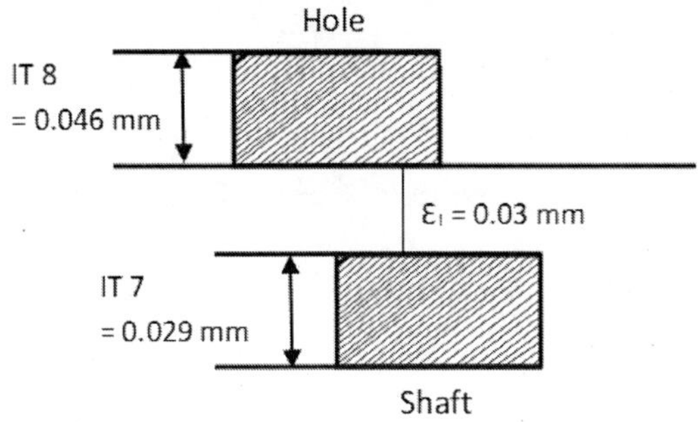

Figure 1.22 Fit diagram of 64 H8f7

IT 7 = 16 i = 29.69 microns = 0.0296 mm

IT 8 = 25 i= 46.4 microns = 0.046 mm

Fundamental deviation for "f" type of shaft

$\varepsilon_f = -5.5\ D^{0.41} = -5.5(63.24)^{0.41} = -30.11$ microns = -0.03 mm

Hole → Max = 64.046 mm, Min = 64.00 mm $64^{+0.046}_{+0.0}$

Shaft → Max = 63.97 mm, Min = 63.941 mm $64^{-0.03}_{-0.059}$

Allowance = 0.03 mm

Example 1.6: Find out and sketch the limits of tolerance and allowance for a 75mm shaft and hole pair nominated H_7S_8. The basic size lies in the range of 50-80 mm. The multipliers for grades 7 and 8 are 16 and 25 correspondingly. The fundamental deviation for 's' shaft is (IT 7 +0.4D) microns.

Solution: $D = \sqrt{50 \times 80} = 63.24$ mm

Fundamental tolerance

$i = 0.45D^{1/3} + 0.001D$

$= 0.45\ (63.24)^{1/3} + 0.001(63.24)$

= 1.85 microns

Hole →Max = 75.029 mm

Min = 75.0 mm

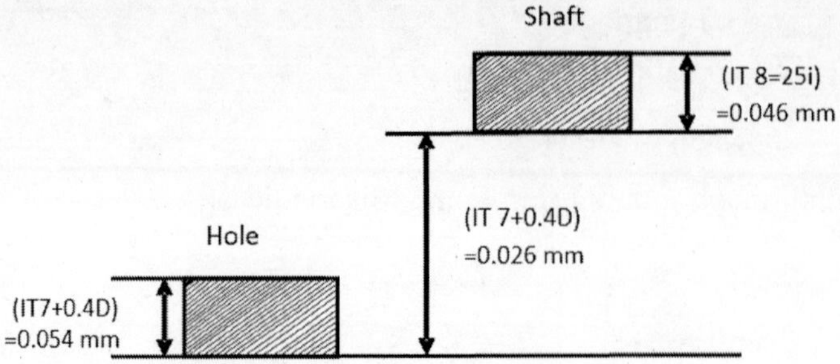

Figure 1.23 Hole and Shaft Tolerance

Shaft → Max = 75.10

Min = 75.054 mm

Allowance = 0.054 + 0.046 = 0.010 mm

Example 1.7: For each of the following shaft and hole pair, calculate shaft tolerance, hole tolerance and analyse whether the pair is,

1. Clearance fit
2. Transition fit
3. Interference fit

Pair 1 : Hole $-\ 50^{+0.50}_{+0.00}$ mm Shaft $-\ 50^{-0.02}_{+0.005}$ mm

Pair 2 : Hole $-\ 50^{+0.25}_{+0.00}$ mm Shaft $-\ 50^{-0.05}_{+0.005}$ mm

Pair 3 : Hole $-\ 50^{+0.04}_{+0.00}$ mm Shaft $-\ 50^{-0.07}_{+0.04}$ mm

Represent the three fits on the same zero line.

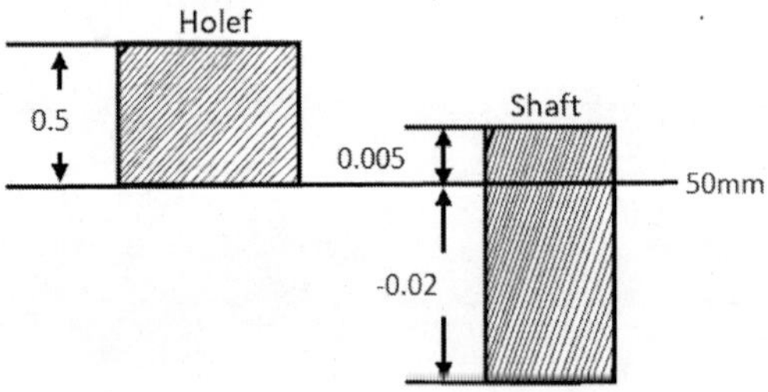

Figure 1.24 Pair 1: Clearance Fit

Hole tolerance = 0.5 mm

Shaft tolerance = 0.005 + 0.02 = 0.025 mm

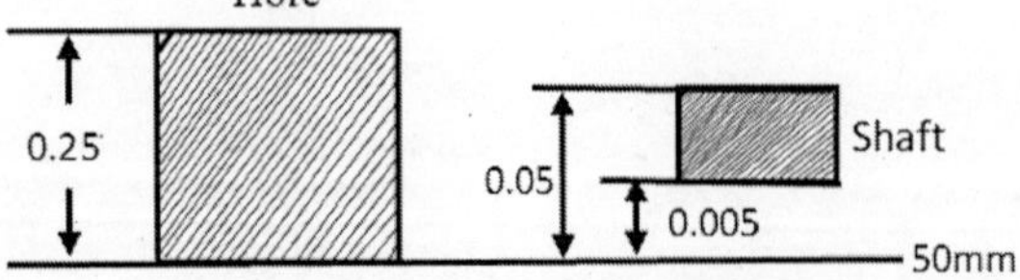

Figure 1.25 Pair 1: Transition Fit

Hole tolerance = 0.25 mm

Shaft tolerance = 0.05 - 0.005 = 0.045 mm

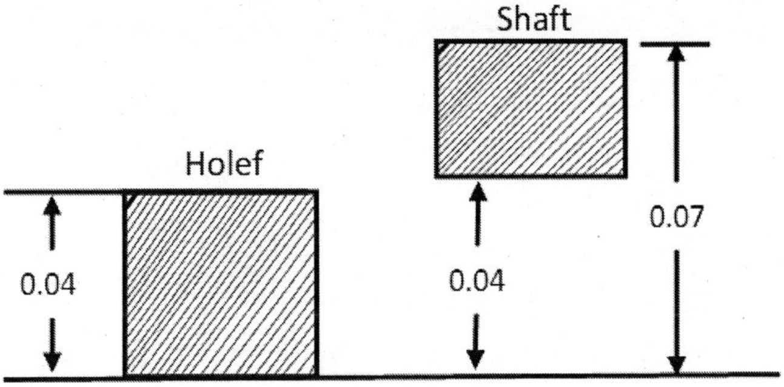

Figure 1.26 Pair 1: Interference Fit

Hole tolerance = 0.04 mm

Shaft tolerance = 0.07 - 0.004 = 0.03 mm

Example 1.8: A 35 mm diameter shaft and bearing are to be assembled with a clearance fit. The tolerance and allowance are as under

Allowance = 0.003 mm

Tolerance on hole = 0.007 mm

Tolerance on shaft = 0.002 mm

Find the limits of size for the hole and shaft if

1. Hole basis system is used
2. Shaft basis system is used

The tolerances are unilaterally discarded.

Solution: Allowance = Minimum clearance

1. **Hole basis system**

Hole → Max = 35.007 mm

Min = 35.0 mm

Shaft → Max = 34.997 mm

Min − 34.995 mm

2. **Shaft basis system**

Hole → Max = 35.01 mm

Min = 35.003 mm

Shaft → Max = 35.0 mm

Min = 34.998 mm

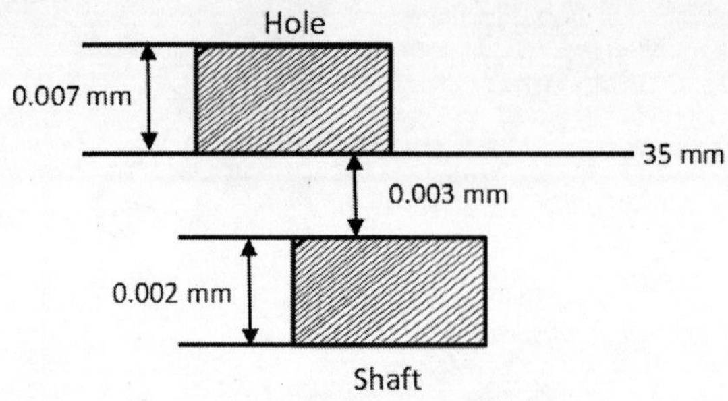

Figure 1.27 Hole Basis

Example: 1.09: The hole and shaft assembly of 30 mm nominal size have tolerances specified as $30^{+0.02}_{+0.00}$ mm for hole and $30^{+0.04}_{+0.07}$ mm for shaft. Determine

1. Maximum and minimum clearance (interface) attainable
2. Allowance
3. Hole and shaft tolerances
4. Fundamental deviation
5. MML for shaft and hole
6. Type of fit.

Solution :

1. Maximum clearance = 0.09 mm

   Minimum clearance = 0.04 mm
2. Allowance = Minimum clearance = 0.04 mm
3. Hole tolerances = 0.02 mm

   Shaft tolerances = 0.03 mm
4. Fundamental deviation of shaft = 0.04 mm
5. Maximum material limit for hole = 30.0 mm

   Maximum material limit for shaft = 29.96 mm
6. Clearance fit

1.31 Need of Limit Gauges

Absolute measurement is a time taking process and a technical worker is required to carryout the aforesaid task. This makes the inspection expensive and hence the cost of product. Gauges (GO or NO GO) are designed to check both holes and shafts that whether they come within their tolerance limits or not. But if the job is very precise then the absolute measurement is done and not gauged even if there are so many components.

When the basic size increases beyond a certain value GO and NO GO are provided on a separate handle [Fig. 1.28].

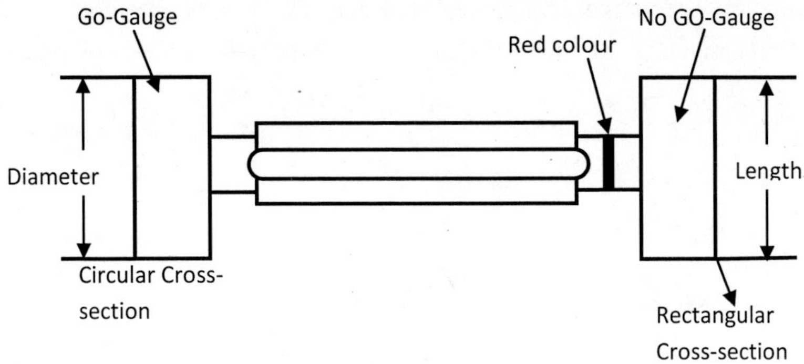

Figure 1.28 Go and No Go Gauge

Since it is not possible to make anything without tolerance, so tolerances are also provided on the gauges and it is called necessary evil. Gauge tolerance is 1/10th of work tolerance. These gauges are considered according to Taylors principle in simple language which has two statements:

1. GO and NO GO gauges must be designed to check size as well as shape and shall be in the full form i.e. the full working length of hole or shaft.
2. NO GO gauges are designed to check one element at a time. e.g. suppose there is a hole which became oval during machining which one cannot see with naked eyes.

1.31.1 Work Shop Gauges:

The philosophy of workshop gauge is that we will not produce anything defective and hence the tolerance on gauges will be towards the work tolerance. For hole there will be plug gauge and since it is repeatedly moving inside the hole, its diameter will keep on decreasing. So direction of wear will be in the downward direction. Ring gauge moves over the shaft so its diameter will keep on increasing so the direction of wear will be in the upward direction. A gauge can have dimension anywhere in tolerance zone. So when either hole or shaft which is within the tolerance limits near the boundaries inspected by these gauges it will be rejected. So by using these gauges some of the good parts will be treated as bad parts. If the GO gauge fits into the hole, it shows that the hole's dimension is greater than the lower limit. If the NO GO gauge does not fit, it indicates that the size of the hole is smaller than the maximum limit. If both GO and NO GO gauges goes inside the hole it means the hole is oversize and when ever GO gauges does not go it means hole is undersize. By repeatedly using GO gauge its dimension changes. GO gauge is continuously being used until wear limit which is 4u's is reached. Once GO gauge reaches wear limit it is thrown away and a new gauge is taken.

1.31.2 Inspection Gauges:

The philosophy of inspection department is that already lot of effort has been taken to make the product so no good part should be rejected. So the tolerance of the gauges is away from the work tolerance as shown in Figure 1.29. It can be seen that when holes and shafts comes slightly beyond the tolerance zones, it will be accepted by the gauges. So some of the bad parts will be treated as good parts by these gauges. Since inspection gauges are already inaccurate so no wear allowance is given to the GO gauges.

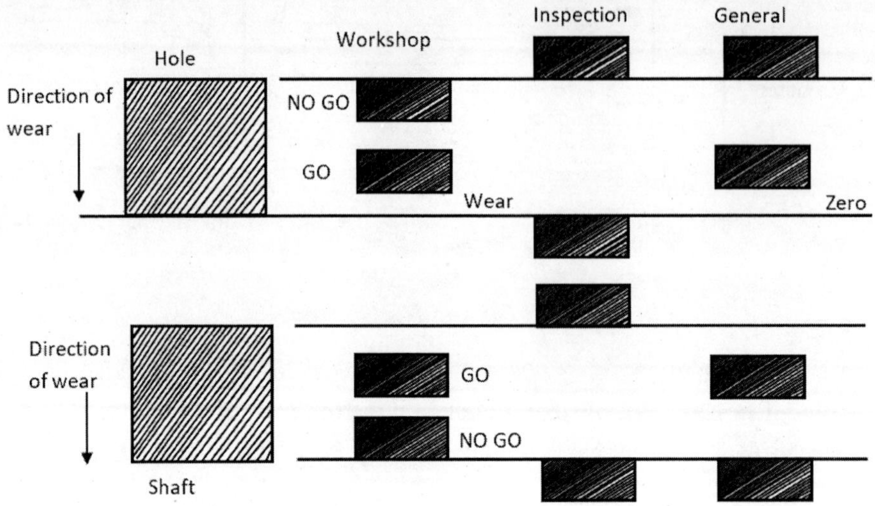

Figure 1.29 Tolerance of Gauges

1.31.3 General Purpose (ISO) Gauge

According to ISO: 2709 there are no separate gauge for workshop and inspection department. For general purpose gauge GO gauge is taken from workshop type and NO GO is taken from inspection gauges furthermore, Slip Gauge, Angle block, and Sine Bar can be employed for measurements. However, the comprehensive details of these instruments beyond the coverage of this book.

1.32 Workpiece control

Workpiece control refers to the process of ensuring that the workpiece is correctly positioned relative to the tool during manufacturing or machining operations. This ensures that the desired dimensions are achieved without being affected by factors such as tool contour, size, or previous processes.

- Workpiece control deals only with the dimensions that need to be obtained, ensuring that they are achieved accurately.
- It does not influence the dimensions created by the tool's contour or size, meaning that the tool's characteristics do not impact the desired dimensions of the workpiece.
- It also does not affect the dimensions produced by previous processes, maintaining consistency and accuracy throughout the manufacturing or machining workflow.

The essence of workpiece control can be summarized as:

- Consistently positioning the workpiece in relation to the tool despite variations or variables in the process.
- Holding the workpiece in the desired position against the forces exerted by the tool during machining.
- Restricting any deflection or deviation of the workpiece from its intended position during the machining process.

1.33 Workpiece Control Theories

Workpiece control theories encompass various principles and methodologies aimed at achieving accurate and consistent positioning of the workpiece relative to the tool during manufacturing or machining processes. Some theories are as follows

1. Equilibrium theory
2. Location theory
3. Geometric control
4. Dimensional control
5. Mechanical control

1.33.1 Equilibrium Theory

Equilibrium theory in workpiece control encompasses both linear equilibrium and rotational equilibrium principles to ensure stability and accuracy during machining operations.

1. **Linear Equilibrium:** Linear equilibrium focuses on maintaining balance along the linear axes (X, Y, and Z) during machining processes. It involves controlling forces acting in these directions to prevent undesired movement or deflection of the workpiece.

   Key considerations for achieving linear equilibrium include:

   - Balancing cutting forces: Ensuring that the cutting forces exerted by the tool are adequately countered by clamping or holding forces to prevent workpiece displacement.
   - Rigidity of the machining setup: Using sturdy fixtures, workholding devices, and machine tools to minimize flexing or deformation of the workpiece or machine components.
   - Minimizing external disturbances: Mitigating factors such as vibrations, thermal expansion, and external forces that could disrupt linear equilibrium and affect machining accuracy.

2. **Rotational Equilibrium:** Rotational equilibrium pertains to maintaining stability around rotational axes (such as pitch, yaw, and roll) during machining processes, especially in operations like milling and turning. Achieving rotational equilibrium involves:

   - **Balancing torque forces**: Ensuring that torque generated by cutting operations is counteracted by the rigidity of the work holding system to prevent rotational movement or misalignment of the workpiece.
   - **Proper tool geometry and alignment:** Selecting cutting tools with appropriate geometry and ensuring correct tool alignment to minimize the tendency for the workpiece to rotate or twist during machining.
   - **Controlling chatter and vibrations**: Implementing strategies to dampen or suppress chatter and vibrations that can induce rotational instability and compromise machining accuracy.

By addressing both linear and rotational equilibrium aspects, workpiece control can effectively maintain stability and positional accuracy throughout machining processes, resulting in precise and consistent dimensional outcomes for manufactured parts.

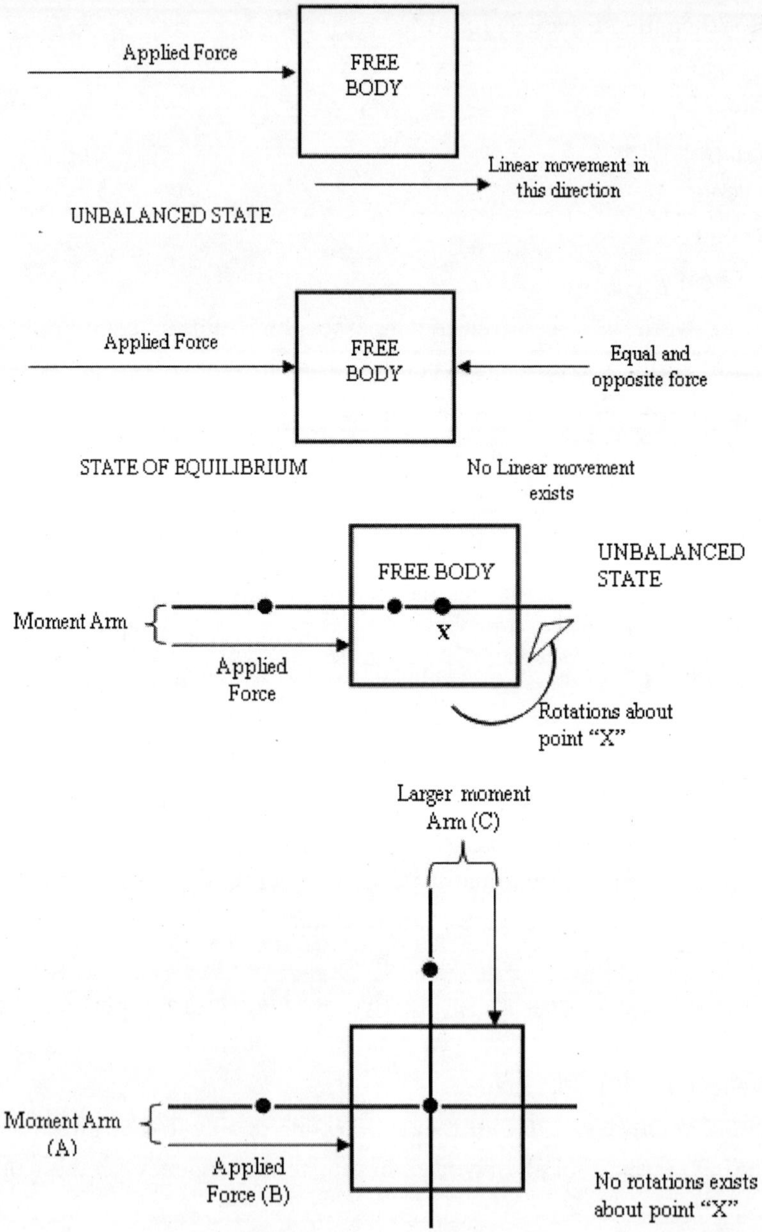

Figure 1.30 Free Body Diagram for Linear and Rotational Equilibrium

1.33.2 Location Theory

Location theory deals with the optimal placement and arrangement of locators or fixturing elements to securely position and hold workpieces during machining or assembly processes. The 3-2-1 location system is a commonly used method for precisely positioning workpieces in three-dimensional space.

3-2-1 Location System

The 3-2-1 location system provides a simple and effective way to accurately locate and hold a workpiece in a specific orientation using three locators on mutually perpendicular sides. The "3-2-1" notation refers to the number of locators placed on each axis: three on one axis, two on the second axis perpendicular to the first, and one on the third axis perpendicular to the other two.

- **Locators:** These are positioned along one axis to restrict movement in two directions while allowing movement along the third axis. Typically, these locators are placed on the base or fixture surface.
- **Locators:** Placed perpendicular to the first set of locators, these restrict movement in one direction while allowing movement along the remaining two axes.
- **1 Locator:** Positioned perpendicular to both the first and second sets of locators, this restricts movement along one axis while allowing freedom of movement along the other two axes.

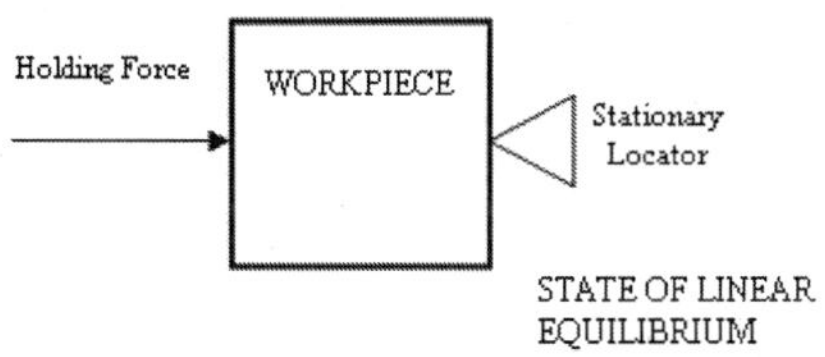

Figure 1.31 Equilibrium in the Workpiece Holder

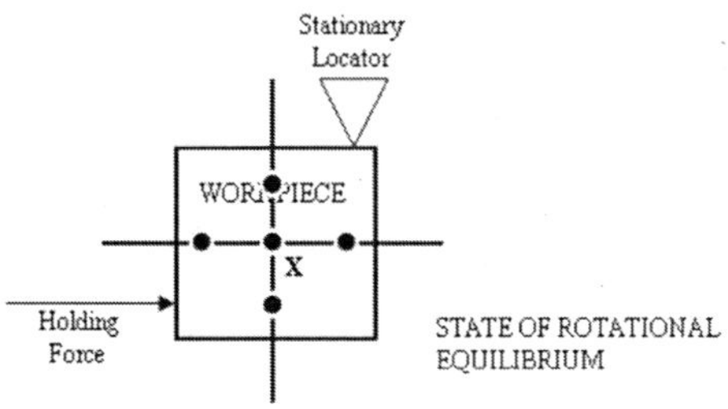

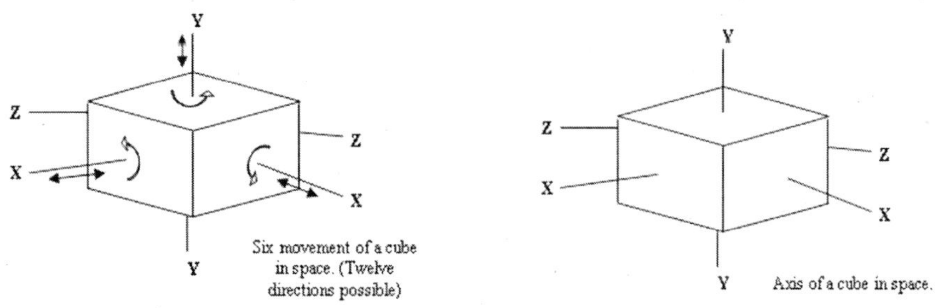

Figure 1.32 Movements in Space

By strategically arranging locators in this manner, the workpiece can be securely held and accurately positioned for machining or assembly operations while ensuring proper alignment and orientation.

1.33.3 Geometric control

Geometric control aims to maintain the stability of the workpiece during machining or assembly operations. This stability ensures that the workpiece remains securely positioned and does not shift or move unexpectedly, leading to accurate and consistent results.

Factors contributing to an unstable workpiece in a location system:

1. **Too close placement of locators:** Placing locators too close to each other can limit the contact area with the workpiece, reducing stability.
2. **A top-heavy workpiece:** Workpieces with a high center of mass or uneven weight distribution are prone to tipping or rocking away from the locators.
3. **Poor placement of holding force:** Inadequate or uneven application of holding forces, such as clamping forces, can result in an unstable workpiece.
4. **Insufficient number of locators:** A lack of sufficient locators may not provide adequate support to prevent workpiece movement.

1.34 Advantages of good geometrical control

1. **Automatic resting of the workpiece against locators:** Well-designed locators facilitate automatic alignment and positioning of the workpiece, reducing the reliance on operator skill and effort.
2. **Reduced workpiece shift tendency:** Properly distributed locators and holding forces minimize the tendency for the workpiece to shift or move away from its intended position during machining or assembly.
3. **Reduced effect on workpiece position**: By providing widespread locators, the impact of external factors such as dirt, chip particles, or surface irregularities on the workpiece position is minimized, enhancing stability and accuracy.

To enhance understanding, let us consider an example of applying geometric control to rectangular shapes using the 3-2-1 placement method.

Achieving geometric control on rectangular workpieces using the 3-2-1 location method involves strategically placing locators according to specific rules. Here's how you can implement these rules:

1.34.1 Placement on Largest Surface

Rule (i): Begin by placing three locators on one of the largest surfaces of the rectangular workpiece. These locators determine the plane of the workpiece position. The selection of this surface is typically based on the orientation of the part within the machining setup and the accessibility for fixture design.

1.34.2 Placement on Second Largest Surface

Rule (ii): Next, place two locators on one of the second largest surfaces, usually an edge of the rectangular workpiece. This provides additional support and stability to the workpiece, particularly against translational movements along the plane defined by the first set of locators.

1.34.3 Placement on Smallest Surface

Rule (iii): Finally, place one locator on the smallest surface of the rectangular workpiece, typically an end. This locator serves to further constrain the workpiece and prevent rotational movements, particularly around axes perpendicular to the plane defined by the first set of locators.

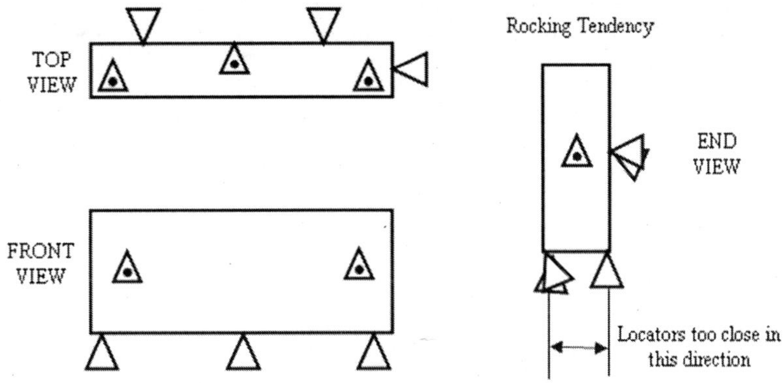

Figure 1.33 Geometric Control of a Rectangular Shape

1.35 Dimensional Control

- Dimensional control focuses on meeting the specified dimensions outlined on the part print.
- It does not encompass other aspects such as surface finish, heat treatment, chemical content, or testing, which are covered separately in the manufacturing process.
- Dimensional control is distinct from controlling specifications. While specifications define the quality standards and requirements for the product, dimensional control specifically addresses achieving the specified dimensions.
- **Preference over Geometric Control:** In cases where it's challenging to achieve both geometric and dimensional control simultaneously due to the dimensioning method, dimensional control takes precedence. This preference is due to the criticality of meeting dimensional requirements for the functionality and fit of the part.
- **Requirements for Achieving Dimensional Control:** Achieving good dimensional control may require higher operator skill or the use of more advanced and expensive tooling. Process engineers may collaborate with product engineers to make necessary adjustments to the part print to facilitate better dimensional control.
- **Achieving Good Dimensional Control:** Correct selection of the surface for placing locators is crucial, as it influences the stability and accuracy of workpiece positioning. Proper placement

of locators on the selected surface ensures that the workpiece is securely held in the desired position, minimizing dimensional variations during machining.

- Certain dimensions on the workpiece must be maintained with greater precision than what is provided. Exceeding the stipulated tolerances would lead to increased production expenses. Therefore, ensuring dimensional control is the most cost-effective method for producing measurements that fall within specified tolerances.
- For optimal dimensional control, it is advisable to position locators on either of the two surfaces that display the dimension on the part print.

By emphasizing proper surface selection and locator placement, process engineers can effectively achieve good dimensional control, ensuring that the manufactured parts meet the specified dimensional requirements.

1.36 Mechanical Control

Mechanical workpiece control entails the use of physical mechanisms and systems to guarantee the secure positioning, orientation, and stability of workpieces during machining operations. This method depends on a range of mechanical concepts and components to attain accurate control over the workpiece.

The following methods are implemented in mechanical control.

1.36.1 Fixturing and Clamping

- Fixturing involves the use of specialized devices called fixtures to hold the workpiece securely in place during machining.
- Clamping mechanisms, such as vises, clamps, or bolts, apply force to hold the workpiece firmly against the fixture or machine table.
- The design and selection of fixtures and clamping methods depend on the specific geometry, material, and machining requirements of the workpiece.

1.36.2 Locating and Positioning

- Locating features, such as pins, holes, or surfaces, are used to precisely position the workpiece within the fixture or on the machine tool.
- Mechanical stops, blocks, or adjustable components are employed to limit the movement of the workpiece and ensure consistent positioning for accurate machining operations.

1.36.3 Alignment and Orientation

- Mechanical alignment methods are used to ensure that the workpiece is properly aligned with respect to the machining tool or fixture.
- This may involve using jigs, wedges, shims, or adjustable components to align the workpiece to specific angles or orientations required for machining.

1.36.4 Flexibility and Adaptability

- Mechanical control systems should be designed with flexibility and adaptability to accommodate variations in workpiece geometry, size, and material.
- Adjustable fixtures, modular components, and quick-change mechanisms enable efficient setup and reconfiguration for different machining tasks.

1.36.5 Operator Skill and Monitoring

- Skilled operators are required to interpret drawings, set up machines, and adjust fixtures effectively.
- Continuous monitoring of machining processes is essential to detect any deviations from desired dimensions or tolerances, allowing for timely adjustments to maintain control.

Manufacturers can attain accurate and dependable workpiece control by utilizing these mechanical control methods. This leads to the production of high-quality machined parts that conform to the necessary standards and tolerances.

UNIT SUMMARY

In order to improve the quality, it is essential to possess a thorough comprehension of the procedure. This Unit provides a thorough introduction to the notion of process and highlights the importance of establishing a mentality focused on processes. Throughout time, a multitude of methodologies and ideologies have been created to efficiently govern and improve processes. Thorough quality assurance. Lean and Six Sigma stand out as particularly noteworthy among these approaches. Hence, it is important to have a comprehensive grasp of these concepts in order to efficiently coordinate efforts to improve processes. The second portion of this Unit delves into the concepts of Dimensional analysis, Tolerance analysis, and work piece control. These strategies are frequently employed in engineering to streamline issue descriptions by reducing the number of variables and facilitating more logical deductions.

EXERCISES NO: -1

Multiple Choice Questions

1. What does equilibrium theory in workpiece control primarily focus on?
    a) Ensuring accurate location of the workpiece
    b) Achieving tight geometric tolerances
    c) Maintaining stability and balance of the workpiece
    d) Controlling dimensional variations
2. Which of the following is a key aspect of location theory in workpiece control?
    a) Controlling surface roughness
    b) Ensuring proper alignment of cutting tools
    c) Maintaining dimensional accuracy
    d) Accurately positioning the workpiece relative to the machining equipment

3. Which of the following is a key aspect of dimensional control in workpiece manufacturing?
   a) Minimizing workpiece vibration
   b) Ensuring tight geometric tolerances
   c) Controlling cutting forces
   d) Maintaining specified tolerances on workpiece dimensions
4. What is the primary function of process planning in manufacturing?
   a) Designing product prototypes
   b) Translating design specifications into a detailed manufacturing plan
   c) Marketing the product to potential customers
   d) Managing inventory levels
5. What are the factors affecting process planning?
   a. Material properties and characteristics
   b. Marketing strategies
   c. Employee satisfaction levels
   d. Office location
6. What does plant capacity refer to in process planning?
   a. The maximum number of employees a plant can hire
   b. The maximum output a manufacturing facility can produce within a given time frame
   c. The number of plants owned by a manufacturing company
   d. The size of the plant's cafeteria
7. What is machine capacity in process planning?
   a. The maximum number of machines in a plant
   b. The number of machines currently operational in a plant
   c. The maximum production capability of individual machines or equipment
   d. The size of the machines in a plant
8. What does preliminary part print analysis involve?
   a. Examining engineering drawings to understand design requirements before developing the manufacturing process
   b. Analyzing market trends to determine product demand
   c. Testing prototypes for product performance
   d. Conducting employee training sessions on manufacturing processes
9. What is the purpose of establishing the general characteristics of the workpiece in preliminary part print analysis?
   a. To determine the cost of manufacturing
   b. To identify key features, dimensions, and tolerances specified in the part print
   c. To schedule production timelines
   d. To organize team meetings

10. What is the primary purpose of a Bill of Material (BOM) in Production Planning and Control (PPC)?
   a. To schedule production timelines
   b. To organize employee schedules
   c. To maintain inventory of finished products
   d. To provide a comprehensive list of components and materials required to manufacture a product

11. What information does a BOM typically include?
   a. Employee performance evaluations
   b. Marketing strategies
   c. Component part numbers, descriptions, quantities, and sourcing details
   d. Customer feedback forms

12. What is the purpose of fit and tolerance in engineering?
   a. To determine the weight of the object
   b. To specify the surface finish of the object
   c. To control the mating of parts and ensure proper functionality
   d. To indicate the color of the object

13. What does the term "fit" refer to in engineering?
   a. The ability of two parts to slide past each other
   b. The relationship between two mating parts with respect to their size and shape
   c. The ability of a part to withstand loads without breaking
   d. The process of assembling parts using adhesives

14. Which of the following is NOT a commonly used fit type in engineering?
   a. Clearance fit
   b. Interference fit
   c. Transition fit
   d. Random fit

15. What is the purpose of specifying tolerances in engineering drawings?
   a. To indicate the type of material to be used for the part
   b. To define the allowable variation in dimensions for proper fit and function
   c. To determine the surface finish required for the part
   d. To specify the manufacturing process to be used for the part

Answers of Multiple-Choice Questions

1 c, 2 d, 3 d, 4 b, 5 a, 6 b, 7 c, 8 a, 9 b, 10 d, 11 c, 12 c, 13 b, 14 d, 15 b

Short and Long Answer Type Questions

1. What are the key components typically included in a Bill of Material (BOM)?

2. How does a Bill of Material (BOM) help in preventing overproduction in manufacturing processes?
3. Discuss the role of Production Planning and Control (PPC) in ensuring efficient utilization of resources and timely delivery of products. Provide examples to illustrate your points.
4. Compare and contrast the make or buy decision in production planning, discussing the factors that influence this decision-making process. Provide real-world examples to support your analysis.
5. Define process planning and explain its importance in manufacturing.
6. What are the prerequisites for effective process planning?
7. Identify and explain the factors that can affect process planning decisions.
8. What is the "make or buy" decision in manufacturing? How is it made?
9. Differentiate between plant capacity and machine capacity. How are they important in process planning?
10. Discuss the importance of understanding the nature of the work to be performed in process engineering.
11. What is dimensional analysis, and what are its main objectives?
12. What are equilibrium theories in workpiece control, and how do they impact manufacturing processes?
13. Explain the concept of location in workpiece control. Why is it important?
14. Discuss the principles of geometric control and its role in ensuring product quality.
15. How does dimensional control contribute to maintaining the desired specifications of manufactured parts?

KNOW MORE

REFERENCES

[1] *Spangler, T., R. Mahajan, S. Puckett, and D. Stakem, & quot.; Manual Labor-Advantages, When and Where? & quot.; MSE 427 Term Paper, Lehigh University, 1998.*

[2] *Brown, Emily. & quot.; Advanced Methods in Part Print Analysis for Quality Control. & quot.; International Journal of Mechanical Engineering, vol. 22, no. 4, 2020, pp. 112-125.*

[3] *Chen, Wei, et al. & quot.; An Integrated Approach to Part Print Analysis and Process Optimization & quot.; International Journal of Production Research, vol. 58, no. 2, 2020, pp. 478-491.*

[4] *Hybrid Assembly Lines with Generic Algorithms, Springer Verlag London Limited, UK, 2006. [24]*

[5] *Jones, Matthew, et al. & quot.; Application of Computer Vision in Part Print Analysis for Defect Detection. & quot.; Journal of Manufacturing Systems, vol. 50, 2019, pp. 123-136.*

[6] *Kim, Sung, et al. & quot.;A Review of Dimensional Inspection Techniques for Part Print Analysis .& quot.; Measurement Science and Technology, vol. 28, no. 6, 2017, 065006.*

[7] *Li, Xin, et al. & quot.;Advances in Non-Destructive Evaluation Techniques for Part Print Analysis.& quot.; NDT & E International, vol. 115, 2022, 102420.*

[8] *Patel, Rajesh, et al. & quot.; Automated Part Print Analysis Using Machine Learning Algorithms. & quot.; IEEE*

[9] *Smith, John. & quot.; An Overview of Part Print Analysis Techniques. & quot; Journal of Manufacturing Engineering, vol. 35, no. 2, 2018, pp. 45-58.*

[10] *Sumichrast, R. T., R. R. Rus0seL, and B. W. Taylor, & quot.; A Comparative Analysis of Sequencing for Mixed-Model Assembly Lines in a Just-In-Time Production System, & quot; International Journal of Production Research, Vol. 30, No. 1, 1992, pp. 199-214.*

[11] *Zhang, Li, et al. & quot.; Deep Learning Approaches for Part Print Analysis in Additive Manufacturing. & quot.; Additive Manufacturing, vol. 9, 2021, pp. 75-88.*

-----------------*---------*

2 PRODUCTION FORECASTING & SCHEDULING

UNIT SPECIFICS

The following topics are covered in this unit:

PRODUCTION FORECASTING

Introduction of production forecasting, the strategic role of forecasting in supply chain, Time frame, Demand behavior, Forecasting methods- Qualitative and Quantitative Qualitative and Quantitative, Forecast accuracy.

SCHEDULING

Introduction, Objectives in scheduling, Loading, Sequencing, Monitoring, Advanced Planning and Scheduling Systems, Theory of Constraints, Employee scheduling.

All the topics are discussed in a lucid manner along with their engineering applications. For student's practice, a moderate number of multiple-choice questions, short and long answer type questions and numerical problems are also given.

RATIONALE

Production forecasting is essential in the industry to effectively manage resources, plan production schedules, and meet customer demands efficiently. By forecasting production needs, businesses can strategically align their supply chain processes, ensuring timely delivery of goods while minimizing costs. Understanding demand behavior and employing both qualitative and quantitative forecasting methods enable organizations to make informed decisions about inventory levels, production capacities, and resource allocation. Accurate forecasting not only enhances operational efficiency but also improves customer satisfaction by reducing lead times and preventing stockouts. Scheduling, as a vital aspect of production management, aims to optimize resource utilization, minimize idle time, and enhance productivity. Through loading, sequencing, and monitoring, businesses can achieve better coordination of production activities, ensuring smooth operations and timely delivery of goods. Advanced Planning and Scheduling (APS) systems, along with methodologies like the Theory of Constraints, offer powerful tools for optimizing production schedules and resource allocation. Additionally, effective employee scheduling ensures optimal utilization of workforce resources, further enhancing operational efficiency and organizational performance. In summary, production forecasting and scheduling are indispensable components of industry operations, driving efficiency, reducing costs, and maintaining competitive advantage in today's dynamic marketplace

PRE-REQUISITES

- Fundamental Knowledge of any Manufacturing Industry Layout
- A course on Industrial Engineering
- Fundamental Knowledge of Marketing

UNIT OUTCOMES

After understanding this chapter readers will be able to

U2-O1: Understand the necessity of production forecasting in the manufacturing sector.

U2-O2: Understand many techniques of forecasting methods.

U2-O3: Understand the significance of Production planning and control in the industrial sector.

U2-O4: Understand the process for creating production schedules

U2-O5: Understand various limitations encountered throughout the process of scheduling and strategies to overcome them.

Unit-2 Outcomes	EXPECTED MAPPING WITH COURSE OUTCOMES (1- Weak Correlation; 2- Medium correlation; 3- Strong Correlation)				
	CO-1	**CO-2**	**CO-3**	**CO-4**	**CO-5**
U2-O1	3	3	1	2	0
U2-O2	3	3	1	2	0
U2-O3	3	2	3	3	3
U2-O4	3	2	3	3	3
U2-O5	2	3	3	3	2

2.1 Forecasting and Demand Management

2.2 Introduction

Forecasting serves as a first step to the process of planning. Prior to formulating plans, it is necessary to generate an estimation of the conditions that will be present during a specific period in the future. The process of making estimates and determining their accuracy is crucial, as it is difficult to proceed without some type of estimation.

What is the purpose of forecasting? Forecasting is an essential step in establishing strategies to meet future demand, as there are various circumstances and reasons that make it necessary. The majority of companies are unable to delay their planning until they have actually received orders. Customers typically expect prompt delivery, and producers must proactively forecast future demand for products or services and strategically allocate the necessary capacity and resources to meet that need. Companies producing standardized items must ensure the availability of readily marketable products or, at the absolute least, have materials and subassemblies in stock to save delivery time. Companies that engage in made-to-order production cannot initiate the manufacturing process until a customer place an order. However, they must ensure that they have the necessary manpower and equipment resources readily accessible to fulfill the demand.

Several variables impact the demand for a company's products and services. While it may not be feasible to ascertain all of these characteristics or their impact on demand, it is beneficial to examine a few key elements:

- The general state of the economy and commerce.
- Factors that are competitive.
- Market trends, including fluctuations in demand.
- The organization's own strategies for advertising, promotion, pricing, and product modifications.

2.3 Demand Management

The main objective of a corporation is to cater to the needs and requirements of the consumer. By efficiently catering to clients, the company will frequently achieve success, which financial experts recognize as the primary objective of generating shareholder value. Marketing is concerned with satisfying client requirements, while operations, specifically through materials management, must supply the necessary resources. The synchronization of strategies by these two entities is referred to as demand management.

Demand management is the process of identifying and controlling all requests for products and/or services. It happens in the short, medium, and long term. Long-term demand predictions are essential for strategic company planning, particularly for facilities. Over the medium term, the goal of demand management is to forecast the overall demand for the purpose of planning output. Short-term demand management is necessary for items and is linked to master production schedule. This text primarily focuses on the latter.

In order to efficiently plan material and capacity resources, it is crucial to identify all sources of demand. The entities encompassed in this list consist of both local and international clientele, as well as additional facilities within the same corporate structure, branch warehouses, service parts and requisitions, promotional materials, distribution inventory, and inventory held by customers on a consignment basis. Demand management encompasses various key operations, all of which are primarily influenced by market forces:

- Identifying the entirety of product and service demand within the specified markets. This includes forecasting future events, and it also includes possible market segmentation, client classification, and the identification of demand that adds no value and should be ignored.

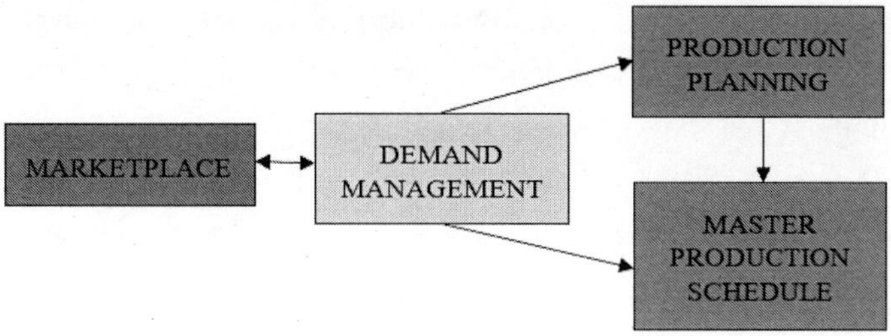

Figure 2.1 The manufacturing planning and control system and demand management

This entails identifying customer preferences regarding the features and designs of existing or prospective future goods or services.

- Analyzing and comprehending all facets of the market that could potentially influence customer demand. This includes factors such as economic conditions, indices, governmental rules and regulations, as well as present or future sources of competition, including the possibility of new competitors.
- Aligning the recognized market demand with the company's capabilities.
- Establishing prioritization for demand in situations where supply is insufficient to meet all demand.
- Committing to delivery guarantees.
- Integration of factory planning and control with the marketplace. Figure 2.1 depicts this correlation through the utilization of a block diagram.
- Processing of orders.

In each of these instances, the process of manufacturing (supply) is being strategically organized to respond to projected demand, as indicated by the forecast.

There is other more tasks or elements involved in demand management. The following items are included:

- Setting and sustaining customer service standards.
- Preparing for new product launches and inventory reduction.
- Managing interplant shipments and distribution needs
- Setting and sustaining inventory targets.
- Setting demand performance parameters and assessing performance. According to Bricks Matter, market-driven proactive demand management involves four main activities: Cecere and Chase: Supply Chains Drive Market Differentiation
- Detecting demand.
- Strategic and market plans shape demand.
- Sales, marketing, and operations use communication, advertising, pricing, and promotion to change demand in company-favorable patterns. This means boosting demand for easy-to-make or profitable products and services and discouraging demand for hard-to-make or unprofitable ones.

As supply chain concepts advance, an additional technique for identifying demand has been created. Collaborative Planning, Forecasting, and Replenishment (CPFR) is an approach that connects trading partners in the supply chain. After that, they create a shared business plan and distribute it to supply chain partners along with sales estimates. More precise demand data can help the specified supply chain. In this closed-loop technique, plans are carried out and then results are examined. The outcomes of analysis can be used to assess and enhance teamwork.

Order processing refers to the handling and management of a customer's order upon its receipt. The product can be sourced either from the existing inventory of completed items or it can be manufactured or assembled based on specific customer orders. When products are sold from inventory, a sales order is generated to authorize the shipment of the goods from inventory. If the product is custom-made or constructed upon request, the sales department is required to create a sales order that clearly outlines the details of the product. If the product is made from conventional components, the assembly process is likely to be straightforward. However, if the product involves substantial engineering, the procedure might become lengthy and difficult. A duplicate of the sales order, which outlines the terms and conditions for accepting the order, is dispatched to the customer. Another duplicate, delivered to the master planner, serves as authorization to proceed with the planning process for manufacturing. The master planner must possess knowledge of the appropriate production items, quantities, and delivery schedules. The sales order must be constructed using language that effectively communicates this information.

2.4 Demand Forecasting

Forecasts are subject to alteration depending on the necessary actions. They must be created for the master production schedule, the sales and operation plan, the strategy plan, and the strategic business plan. Both the business plan and the strategy plan analyze markets and the economic trajectory for a minimum of two to 10 years. They provide you with a sufficient amount of time to strategize for matters that require a significant duration to undergo transformation. In order to ensure smooth production, it is imperative that the business strategy and strategic plan allocate sufficient time for planning resources such as plant growth, procurement of capital equipment, and any other items that require a significant lead time for ordering. The available information is limited, primarily consisting of estimations related to sales units, sales dollars, or capacity. Planning and predictions will be reviewed either quarterly or annually.

Sales and operations planning is responsible for strategizing the production for the upcoming one to three years. In the context of manufacturing, the term refers to the process of producing forecasts on aspects such as financial allocations, workforce arrangements, procurement of items with lengthy lead times, and overall inventory levels. Forecasts are generated for categories or collections of products rather than individual items. It is probable that plans and predictions will be reviewed on a monthly basis. The master production schedule is a comprehensive plan that outlines the production activities scheduled to be completed in the near future, extending several months ahead. The individual item projections are determined by considering many elements such as the master production timetable, inventory levels for each item, the cost of raw materials and parts, labor planning, and other relevant factors. It is probable that plans and predictions will be reviewed on a weekly basis.

2.5 Characteristics of Demand

In this chapter, the term "demand" is used instead of "sales." The differentiation rests in the fact that sales pertain to the tangible things or services that have been acquired, whereas demand represents the particular requirements or wishes stated by the market or clients. At times, it is unfeasible to fulfill the level of demand, leading to sales that fall below the demand. Before exploring the concepts and practices of forecasting, it is recommended to analyze some characteristics of demand that influence the prediction and the specific methods used.

2.6 Demand Patterns

Plotting historical demand data across time will unveil any identifiable patterns or enduring tendencies. A pattern denotes the comprehensive configuration or arrangement of a time series. Although there may be some little deviations, the majority of individual data points tend to cluster closely around the pattern. Figure 2.2 illustrates a hypothetical historical pattern of demand. The graphic illustrates the variability of the real demand over different time periods. There are four variables that are responsible for this phenomenon: trend, seasonality, random fluctuation, and cycle.

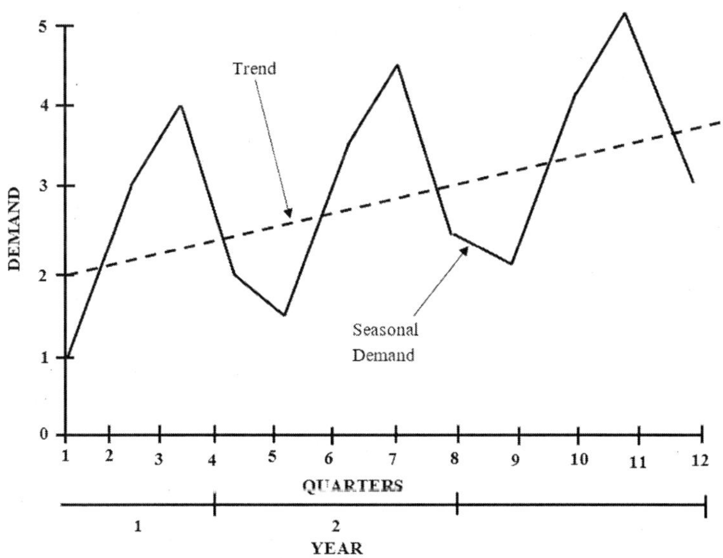

Figure 2.2 Demand as it relates to time

2.6.1 Trend

Figure 2.2 illustrates a consistent and gradual increase in demand over the years. This graph depicts a linear pattern, but it can also exhibit other forms, such as geometric or exponential. The trend can either remain constant, with no variation from one period to another, or it can increase or decrease.

2.6.2 Seasonality

The demand pattern depicted in Figure 2.2 exhibits annual fluctuations that vary depending on the season. This variation could be attributed to meteorological conditions, seasonal festivities, or specific occurrences that occur periodically. Seasonality is commonly associated with annual cycles, although it can also manifest on a weekly or even daily level. The demand for a restaurant fluctuates depending on the time of day, while supermarket sales fluctuate depending on the day of the week.

2.6.3 Random variation

Random variation arises when numerous factors influence demand during specific time periods and happen in an unpredictable manner. The fluctuation can range from little, where real demand closely aligns with the pattern, to substantial, where the sites are widely apart. The pattern of variance can typically be quantified, and this will be addressed in the section on monitoring the forecast.

2.6.4 Cycle

Fluctuations in the economy, occurring over extended periods of time, have a wave-like effect on the demand. Nevertheless, the task of predicting cycles is the responsibility of economists.

2.7 Stable Versus Dynamic

Some products or services exhibit dynamic changes in their demand patterns over time, while others remain static. Objects that maintain a consistent overall form are said to as stable, whereas those that do not are referred to as dynamic. Dynamic fluctuations might impact the pattern, seasonal variations, or unpredictability of the current demand. Forecasting becomes easier as demand stability increases. Figure 2.3 depicts a visual illustration of both stable and dynamic demand. It should be noted that the average demand is consistent for both steady and dynamic patterns. Typically, the projection is made for the typical level of demand.

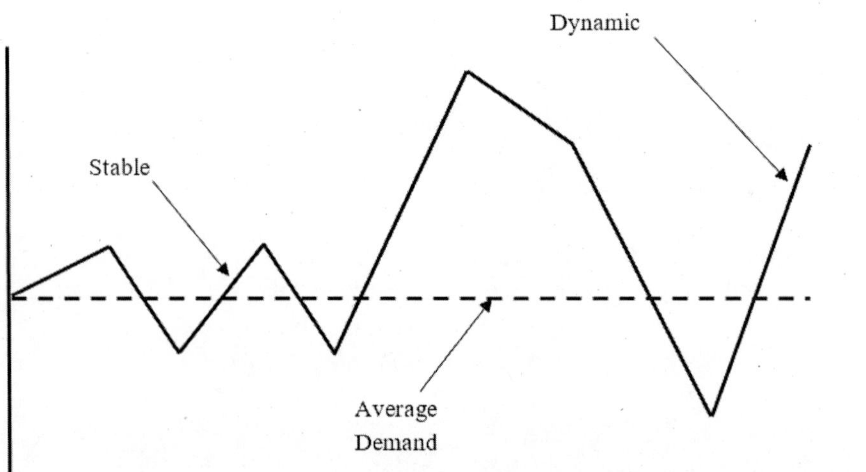

Figure 2.3 Stable and Dynamic Demand

2.8 Dependent Versus Independent Demand

Product or service demand is considered independent when it is not influenced by the demand for any other product or service, or by the internal activities of the organization. Dependent demand refers to the situation when the demand for a certain product or service is generated from the demand for another related one. Dependent demand items do not require a prediction, as their requirements are derived from those of the independent demand item.

Forecasting is only necessary for things that have independent demand. Typically, these items refer to final products or goods that have been completed. However, they should also encompass service parts and things that are provided to other facilities within the same organization (intercompany transfers).

2.9 Principles of Forecasting

Forecasts possess four fundamental traits or principles. Comprehending these concepts will enable the more efficient utilization of predictions. They are straightforward and, to a certain degree, based on common sense.

2.9.1 Forecasts are typically inaccurate

Forecasts endeavor to anticipate the uncertain future and, barring fortuitous circumstances, will inevitably contain some degree of inaccuracy. Mistakes are unavoidable and should be anticipated.

2.9.2 Each forecast must provide a calculation of the margin of error

Given the inherent inaccuracy of projections, the primary concern is determining the extent of their deviation from reality. Each forecast should incorporate an error estimate, typically represented as a percentage (both positive and negative) of the forecast or as a range between the highest and lowest values. Statistical estimates of this mistake can be obtained by analyzing the variability of demand around the mean demand.

2.9.3 Forecasts exhibit higher accuracy when applied to families or groups

The behavior of individual objects inside a group exhibits randomness, even when the group itself possesses highly stable properties. Forecasting the marks of individual students in a class is more challenging than predicting the class average. The high grades are balanced by the low grades. Consequently, predictions are more precise when made for a big collection of items than to when made for individual items within the collection.

2.9.4 Forecasts exhibit higher accuracy for time periods that are closer in proximity

The immediate future is more predictable than the distant future. The majority of individuals' exhibit greater confidence in predicting their activities for the upcoming week compared to making projections for a year ahead. As a previous statement suggests, tomorrow is anticipated to be similar to today.

In the same way, it's easier for a company to guess what demand will be in the near future than in the far future. That is very important for things that take a long time to make, and it's even more important if their demand changes often. It is better to make predictions if wait times are cut down in any way possible.

2.10 Collection and preparation of data

Most forecasts are based on past data that has been changed in some way, either by guesswork or a scientific method. This means that the prediction is only as good as the information it is based on. Three rules about how to collect data are important if you want to get good data.

- Write down the information in the same way that the forecaster needs it. It can be hard to figure out what the forecast is for and why it needs to be made. This is made up of three parts:
  a) If you want to predict how much production will be needed, you need info on demand, not shipments. Shipments show when the things were sent, not always when the customer asked for them. So, supplies don't always really show how much demand there is.
  b) The time frame for the forecast and the plan should match, either in weeks, months, or quarters. If plans happen once a week, the forecast should cover the same amount of time.
  c) The things that are predicted should be the same ones that are used by manufacturing.

For instance, when a product offers multiple options, it is necessary to predict the demand for both the product itself and each individual option. Assume that a company produces a bicycle that is available in three different frame sizes, three different wheel sizes, three different gear changer options (3-speed, 5-speed, or 10-speed), and with or without deluxe trim. There are a total of 54 individual end goods available for sale, which may be calculated by multiplying 3 by 3 by 3 by 2. If each individual scenario were predicted, a total of 54 forecasts would need to be made. A more effective strategy involves predicting both the overall demand and the proportion of that demand that necessitates specific frame sizes, wheel sizes, and other factors. Thus, there would only be a total of 12 forecasts required, consisting of 3 different frame sizes, 3 different wheel sizes, 3 different gear changes, 2 levels of trim, and the bicycle itself. In this example, the duration required to produce the components would be considerably longer when compared to the duration needed to complete a bike. The manufacturing process involves producing components based on component forecasts and subsequently assembling bicycles based on customer orders.

- Write down the details of what happened with the info. Certain events can change demand, and these events should be recorded along with the demand data. Sale events, price changes, weather changes, or. a strike at a competitor's plant are all examples of things that can make demand go up for no reason. It is important that these factors are linked to the past of demand so that they can be added or taken away depending on future circumstances.
- Make separate notes on the desire for each group of customers. A lot of companies get their things to customers in a variety of ways, and each way has its own demand characteristics. One company might sell to several wholesalers who buy small amounts on a regular basis and also to a big store that gets a lot twice a year. It wouldn't help to make predictions about the average demand; instead, each set of needs should be predicted separately

2.11 Forecasting Techniques

There exist numerous forecasting techniques, which may typically be categorized into different groups. Forecasting approaches can be categorized as qualitative or quantitative, and they can rely on extrinsic (external) or intrinsic (internal) elements.

2.11.1 Qualitative Techniques

Qualitative approaches involve making projections based on subjective judgment, intuition, and well-informed opinions. Due to their inherent characteristics, they are subjective. These strategies are employed to predict overall business trends and the potential demand for extensive product categories over a long duration. Consequently, they are primarily utilized by top-level executives. Production and inventory forecasting mostly focuses on predicting the demand for specific end products, and qualitative methods are rarely suitable for this purpose.

When trying to predict the demand for a new product, there is no previous data available to use as a basis for the projection. For these situations, the methods of market research and historical analogy can be employed. Market research is a methodical, structured, and deliberate process used to assess customer opinion or intention. Historical analogy relies on a meticulous examination of the introduction and development of comparable items, with the expectation that the new product will exhibit similar behavior. Another approach is conducting a test-market for a product.

There exist numerous alternative approaches to qualitative forecasting. One method that exemplifies this is the Delphi method. This approach employs a panel of specialists who provide their professional judgments on the probable outcomes.

2.11.2 Quantitative Techniques

Quantitative procedures involve making projections using historical or numerical data, which can come from either within or outside the company.

2.11.3 Extrinsic Techniques

Extrinsic forecasting approaches involve making estimates for a company's product demand based on external variables. Instances of such data include home starts, birth rates, and disposable income. The hypothesis posits that the demand for a particular group of products exhibits a direct proportionality or correlation with the level of activity in another domain. Here are some instances of correlation:

- The sales of bricks are directly linked to the number of housing starts.
- The demand for automotive tires is directly proportional to the level of gasoline use.

Economic indicators encompass housing starts and gasoline usage. They provide a description of the economic conditions that exist during a specific period of time. These signs are commonly referred to be leading indicators since they occur prior to the events they help to predict. For instance, the initiation of housing construction (often indicated by the submission of building permit applications) will result in a demand for various housing components, such as roofing materials, electrical materials, and others. Housing starts serve as a reliable leading indicator for the demand of various housing materials. Several frequently utilized economic indicators include construction contract awards, vehicle output, farm revenue, steel production, and gross national income. Information of this nature is gathered and released by multiple government agencies, financial publications, journals, trade organizations, and banks.

The objective is to identify a metric that is strongly associated with demand and ideally predicts demand, meaning it occurs prior to the actual demand. For instance, the quantity of construction contracts granted during a specific timeframe may influence the sales of building materials in the subsequent timeframe. If a leading indication cannot be found, it may be feasible to utilize a non-leading signal that is forecasted by the government or an organization. Essentially, it involves making a prediction based on a previous prediction.

Extrinsic forecasting is particularly valuable for predicting the overall demand for a company's products or the demand for groups of items. Therefore, it is primarily utilized in business and production planning rather than in the prediction of specific final products.

2.11.4 Intrinsic Techniques

Intrinsic forecasting strategies utilize past data to predict future outcomes. This data is typically documented within the firm and is easily accessible. Intrinsic forecasting methodologies rely on the premise that historical patterns will continue to occur in the future. This assumption has been compared to the act of driving a car while only relying on the rearview mirror. While there is an evident element of truth in this statement, it is equally valid to acknowledge that in the absence of any alternative means of prediction, the most reliable indicator of future outcomes is the analysis of past events.

The importance of intrinsic techniques will be discussed in more detail in the following section, which will focus on a few key techniques. They are commonly used as input for master production scheduling, particularly in situations where the plan's planning horizon requires end item projections.

Some significant intrinsic techniques

Assume for the moment that Table 2.1 shows the monthly demand for a specific item for the prior year. Let's say it's late December and a forecast for the demand in January of next year is needed. There are several guidelines that can be used.:

The demand for this month will remain unchanged compared to the demand for last month. The projected January demand is expected to remain unchanged at 84, which is the same as December. Although it may seem overly simplistic, if there is minimal fluctuation in demand from one month to another, it is likely to be highly functional.

The demand for this month will be equivalent to the demand for the same month last year. The projected demand is expected to be 92, which is same to the demand in January of the previous year. This guideline is sufficient if the demand follows a seasonal pattern and there is minimal upward or downward movement.

Rules that rely on a single month or previous period are not very helpful when there is a significant amount of random variation in demand. Typically, approaches that calculate the average of historical data are superior because they mitigate the impact of random fluctuations.

For instance, the mean of the demand from the previous year can serve as an approximation for the demand in January. A basic mean calculation would not effectively capture fluctuations or variations in the degree of demand. An improved approach would involve utilizing a moving average.

Mean demand This prompts the inquiry of what to predict. As previously mentioned, demand can vary due to stochastic fluctuations. It is more advisable to predict the mean demand rather than speculate on the impact of random fluctuations. The second principle of forecasting, as previously mentioned, states that a forecast should incorporate an estimation of inaccuracy. An estimation can be made for this range, so it is advisable to create a prediction for the average demand and then incorporate the estimated margin of error.

Moving Averages

One easy way to make a prediction is to use the average demand from, say, the last three or six periods as the prediction for the next period. The demand from the most recent period is added to the demand from the previous period at the end of the next period to get a new average that can be used as an estimate. This prediction would always be based on the sum of the real demand over the given time period.

January	92	July	84
February	83	August	81
March	66	September	75
April	74	October	63
May	75	November	91
June	84	December	84

Table 2.1 Monthly Demand History

For instance, let's assume that a decision was made to apply a three-month moving average to the data presented in Figure 2.4. The January forecast, derived from the demand patterns seen in October, November, and December, is as follows:

$$\frac{63 + 91 + 84}{3} = 79$$

Assume that the demand for January was 90 instead of the originally expected 79. The forecast for February will be computed as:

$$\frac{91 + 84 + 90}{3} = 88$$

Example 2.1: The demand for each of the previous three months has been 120, 135 and 114 units, respectively. Using a three-month moving average, calculate the forecast for the fourth month.

Answer: Forecast for month 4 = $\frac{120+135+114}{3} = \frac{369}{3} = 123$

The fourth month's actual demand came out to be 129. Determine the fifth-month forecast.

Forecast for month 5 = $\frac{135+114+129}{3} = 126$

The predicted temperature in the previous conversation was 79 degrees for January and 88 degrees for February. The higher value in January and the lower value in October are indicated by the increasing forecast. A longer time frame (e.g., six months) results in a slower response from the forecast. The shorter the time period used for the moving average, the greater the emphasis placed on the most recent data, resulting in a more rapid response to changes in patterns. Nevertheless, the forecast will consistently fall behind a trend. As an illustration, let's examine the demand history for the previous five periods:

Period	Demand
1	1000
2	2000
3	3000
4	4000
5	5000

There is an increasing trend in demand. The forecast for period 6, using a five-period moving average, is calculated by summing the values of the previous five periods (1000, 2000, 3000, 4000, and 5000) and then dividing the sum by 5, resulting in a forecast of 3000. The projection appears to be somewhat inaccurate as it lags behind the actual demand by a considerable margin. Nevertheless, when employing a three-month moving average, the estimate is calculated by dividing the sum of 3000, 4000, and 5000 by 3, resulting in a value of 4000. Imperfect, however moderately improved. The essence of the matter is that a moving average consistently falls behind a trend, and the extent of this delay increases as more periods are incorporated into the average. If, on the other hand, there is no trend and real demand changes a lot because of random factors, then a moving average based on a few periods responds to the changes instead of predicting the average. Take a look at this history of demand:

Period	Demand
1	2000
2	5000
3	3000
4	1000
5	4000

It's random and doesn't follow a trend. If you look at the five-month moving average, you can see that the next month will see 3000. All the numbers are shown here. If you look at the average of the last two months, here are the predictions for the next three, four, five, and six months:

Forecast for third month = (2000+ 5000) ÷ 2 = 3500

Forecast for fourth month= (5000+3000) ÷ 2 = 4000

Forecast for fifth month = (3000+1000) ÷ 2= 2000

Forecast for sixth month (1000+ 4000) ÷ 2 = 2500

The prediction is not stable because it is based on a two-month moving average, which changes very quickly based on new demand.

Moving averages are best for predicting the demand for stable goods where there isn't much of a trend or variation. You can also get rid of random changes with moving averages. This makes sense because there are times when demand is high and times when demand is low. This is because overall market demand stays the same and people often buy things ahead of time because of sales or outside factors like the weather or holidays. If you buy now, future sales will be lower.

One problem with moving averages is that you have to keep track of several time periods for each thing you want to predict. It will take a lot of computer space or office work to do this. The equations are also hard to do. Exponential smoothing is a popular way to make predictions. It works the same way as a moving average, but it doesn't need to keep as much data and is easier to calculate.

2.11.5 Exponential Smoothing

Exponential smoothing eliminates the need to retain months of historical data in order to generate a moving average. This is because the previously computed forecast already takes into account this historical information. Thus, the forecast can be derived from the previous computed forecast and the updated data.

Based on the information provided in Figure 2.4, let's assume that the average demand of the past six months (80 units) is utilized to predict the demand for January. If the actual demand at the end of January is 90 units, the prediction for July is disregarded and the new projection is determined based on the demand in January. By calculating the average of the previous prediction (80) and the actual demand for January (90), the new forecast for February is determined to be 85 units. This formula assigns equal importance to the most recent month and all preceding months in the forecast. If this does not appear appropriate, a reduced emphasis could be placed on the most recent actual demand and a greater emphasis on the previous projection. It may be more advantageous to allocate only 10% of the weight to the most recent month's demand and 90% of the weight to the previous projection. Under those circumstances, February forecast = 0.1(90) + 0.9(80) = 81

It is important to observe that this projection did not increase as significantly as the prior computation, when equal importance was given to both the previous forecast and the most recent actual demand. An advantage of exponential smoothing is its flexibility in assigning weights to new data.

The weighting assigned to the most recent observed demand is referred to as a smoothing constant and is symbolized by the Greek letter alpha (α). The value is consistently represented as a decimal and normally falls within the range of 0 to 0.3.

Typically, the equation for determining the updated prediction is as follows:

New forecast = (α) (latest demand) + (1 - α) (previous forecast)

Example 2.2: While the actual demand for May was 190, the initial forecast for May was 220. Calculate the June forecast, given that alpha is 0.15. Calculate the anticipated demand for July given that the demand in June was 218.

Answer: June forecast = (0.15) (190) + (1 -0.15) (220) = 215.5

July forecast = (0.15) (218) + (0.85) (215.5) 215.9

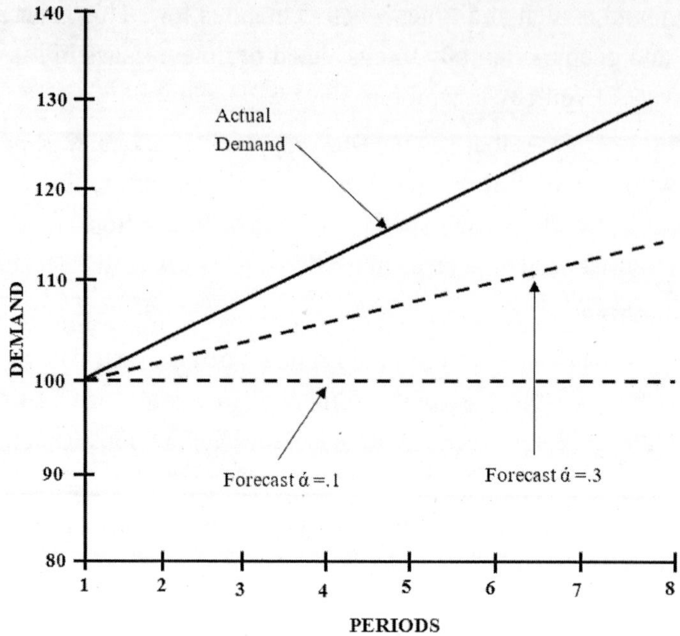

Figure 2.4 Exponential forecast where trend exists

Using exponential smoothing is a regular way to keep item predictions up to date. For things that are stable, it works pretty well. In general, it has been found to be good for making short-term predictions. It's not good enough when demand is low or changes often.

Exponential smoothing can find trends, but if there is a clear trend, the forecast will be later than real demand. There is a positive trend in Figure 2.4, which shows a graph of the exponentially smoothed forecast that is behind the real demand. Take note that the forecast with the bigger an is more in line with real demand.

A little trickier method called "double exponential smoothing" can be used if there is a trend. This method works the same way, but it checks a trend line to see if each new number of the forecast is going up or down. We can't talk about double exponential smoothing in this book.

There is a problem with picking the "best" alpha number. If a small number like 0.1 is used, the old prediction will be given more weight, and new trends will not be noticed as quickly as could be wanted. If a bigger number like 0.4 is used, the forecast will change quickly when demand changes, and it will be all over the place if there is a big random change. PC exercise is a good way to find the best alpha factor. Forecasts are made with different alpha factors based on real past demand for certain products to see which one fits the pattern of demand the best.

2.11.6 Seasonality

Several products exhibit a seasonal or periodic pattern of demand, such as skis, lawnmowers, bathing suits, and Christmas tree lights. Products with demand that fluctuates based on the time of day, week, or month are not as readily apparent. Examples of these could include diurnal electricity consumption or weekly grocery purchasing. Power consumption reaches its highest levels between 4:00 P.M. and 7:00 P.M., whereas stores see their highest levels of activity at the end of the week or before to specific holidays.

2.11.6.1 Seasonal Index

The seasonal index is a valuable measure of the extent of seasonal fluctuations in a product. This index provides an approximation of the extent to which the demand for the product throughout the season will deviate from the typical demand. For instance, the average demand for swimsuits might be 100 every month. However, in July, the average demand could increase to 175, while in September, it could decrease to 35. The July demand index is 1.75, whereas the September demand index is 0.35.

Year	Quarter				
	1	2	3	4	Total
1	122	108	81	90	401
2	130	100	73	96	399
3	132	98	71	99	400
Average	128	102	75	95	400

Table 2.2 Seasonal Index

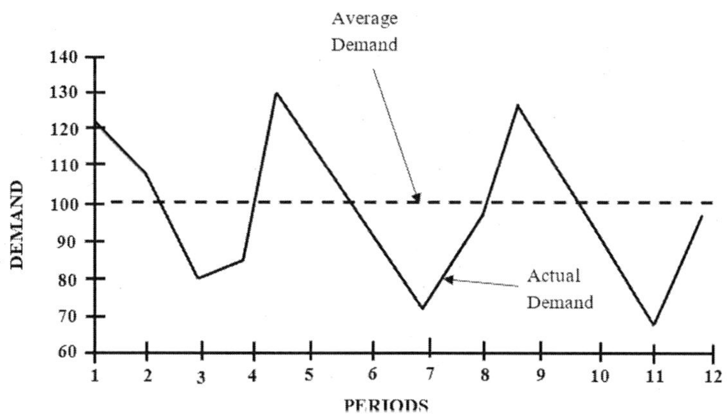

Figure 2.5 Seasonal Sale

The formula for the seasonal index is as follows:

$$\text{Seasonal index} = \frac{\text{period average demand}}{\text{average demand for all periods}}$$

Depending on what underlies the seasonal patterns in demand, the period's frequency could be anywhere from daily to weekly to monthly or quarterly.

The average demand for all periods is a metric that mitigates the impact of seasonality through averaging. This is referred to as the seasonally adjusted demand. The preceding equation can be restated in the following manner:

$$\text{Deseasonalized demand} = \frac{\text{period average demand}}{\text{seasonal index}}$$

Example 2.3: Figure 2.6 displays a product that is influenced by seasonal fluctuations in demand on a quarterly basis, as well as the demand patterns observed over the past three years. While there is no discernible pattern across time, there is a clear and consistent seasonal variation. The mean quarterly demand is 100 units. Figure 2.6 depicts a graph illustrating both the real seasonal demand and the average quarterly demand. The average demand displayed is the historical average demand across all time periods. Please note that this projection pertains to the average demand, not the seasonal demand.

Answer: The seasonal indices can now be calculated as follows:

$$Seasonal\ index = \frac{128}{100} = 1.28\ (quarter\ 1)$$

$$= \frac{102}{100} = 1.02\ (quarter\ 2)$$

$$= \frac{75}{100} = 0.75\ (quarter\ 3)$$

$$= \frac{95}{100} = 0.95\ (quarter\ 4)$$

Total of seasonal indices = 4.00

It is significant to note that the number of periods equals the sum of the seasonal indices. This method is an effective means of verifying the accuracy of the calculations.

2.11.6.2 Seasonal Forecasts

It is important to observe that the sum of all the seasonal indices is equal to the entire number of periods. This method is an effective means of verifying the accuracy of the calculations.

Seasonal demand = (seasonal index) (average demand)

$$\text{Deseasonalized demand} = \frac{\text{actual seasonal demand}}{\text{seasonal index}}$$

Example 2.4: The company in the last problem thinks that they will need 420 units each year next year. Figure out what you think the sales will be each quarter.

Answer

- Expected quarter demand = (seasonal index) × (average quarterly demand)
- Expected first - quarter demand = 1.28 x 105 = 134.4 units
- Expected second - quarter demand = 1.02 x 105 = 107.1 units
- Expected third - quarter demand = 0.75 x 105 = 78.75 unit

- Expected fourth - quarter demand = 0.95 x 105 = 99.75 units
- Total forecast demand = 420 units

2.11.6.3 Deseasonalized Demand

Forecasts don't take chance changes into account. They are made for average demand, and seasonal demand is found by taking the average and adding up the monthly indices. Figure 2.7 shows both the real demand and the average demand that was predicted. The deseasonalized demand is the same thing as the expected average demand. Historical data shows real seasonal demand, and it needs to be "deseasonalized" before it can be used to make an average demand prediction.

It's also not very useful to compare sales from different times of the year unless deseasonalized data is used. One business that sells tennis rackets has found that summer is their busiest time. However, some people play tennis inside, so there is still demand in the winter. How could you compare January demand to June demand if demand was 5,200 units in January and 24,000 units in June?

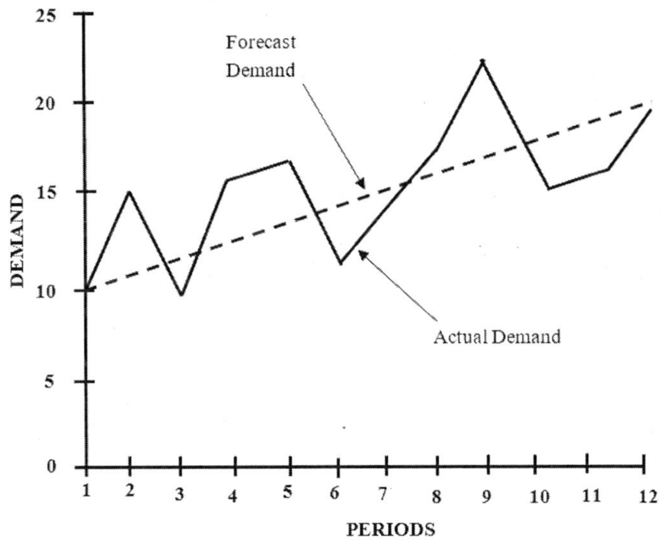

Figure 2.6 Seasonal Demand

Which month had more demand? If there is seasonality, it wouldn't make sense to compare real demand. The data needs to be deseasonalized so that it can be compared.

From the earlier seasonal equation, we can get the following equation to figure out deseasonalized demand for each time period:

$$\text{Deseasonalizeddemand} = \frac{\text{Actual Seasonal Demand}}{\text{Seasonal Index}}$$

Example 2.5: The demand for tennis rackets in January is 5200 units, whereas in June it increases to 24,000 units. Compute the deseasonalized demand for January and June given the seasonal indices of 0.5 and 2.5, respectively. What is the comparison between the two months?

Answer: Deseasonalized January demand = 5200 0.5 ÷ 0.5 = 10,400 units

Deseasonalized June demand = 24,000 ÷ 2.5 = 9600 units

It is now possible to compare the levels of demand in June and January. When comparing the demand levels of January and June after removing seasonal variations, it is evident that the demand in January is higher than the demand in June.

Deseasonalized data must be used in forecasting. The average demand serves as the basis for forecast generation, and the seasonal demand forecast is obtained by multiplying the average demand by the appropriate season index.

The guidelines for making predictions with seasonality are as follows:

- Utilize exclusively deseasonalized data for the purpose of making forecasts.
- Provide a prediction of demand that has been adjusted to remove seasonal variations, namely focusing on the base projection rather than the seasonal demand.
- Determine the seasonal forecast by utilizing the seasonal index and applying it to the base prediction.

Example 2.6: The corporation use exponential smoothing as a method to predict the demand for its products. In April, the projection adjusted for seasonal variations was 1000 units, whereas the actual demand during that season was 1250 units. The April seasonal index is 1.2, while the May seasonal index is 0.7. Given that the value of an is 0.1, compute the following:

a. The deseasonalized actual demand for April.
b. The deseasonalized May forecast.
c. The seasonal forecast for May.

Answer

a) Deseasonalized actual demand for April = $\frac{1250}{1.2} = 1042$

b) Deseasonalized May forecast = ά (latest actual) + (1 - ά) (previous forecast)

i. = 0.1(1042) + 0.9(1000) = 1004

c) Seasonalized May forecast = (seasonal index) (deseasonalized forecast)

i. = 0.7(1004) = 703

2.12 Forecast Tracking

Forecasts are frequently inaccurate, as was mentioned in the section on forecasting principles. This can be attributed to a number of factors, including consumer and economic behavior as well as human involvement. If it were possible to gauge a forecast's accuracy, forecasting techniques could be enhanced and more accurate estimates with error correction. The forecast needs to be monitored since it makes no sense to carry out a plan based on shoddy forecast data. Comparing the actual demand with the forecast is the process of tracking the forecast.

2.13 Forecast Error

This is the difference between what actually happened and what was expected to happen. There are two types of error: bias and chance variation.

2.13.1 Bias

Cumulative actual demand may not be the same as forecast. Examine the data presented in Table 2.3. The actual demand deviates from the projection, and over the course of six months, the cumulative demand exceeds the projected value by 120 units.

Bias exists when there is a discrepancy between the cumulative actual demand and the cumulative forecast. This indicates that the projected average demand has been inaccurate. In the given case depicted in Table 2.3, the predicted average demand was 100 units, however the actual average demand turned out to be 120 units, calculated by dividing 720 units by 6. Figure 2.7 displays a graph representing the cumulative projected and actual demand.

Month	Forecast		Actual	
	Monthly	Cumulative	Monthly	Cumulative
1	100	100	110	110
2	100	200	125	2235
3	100	300	120	355
4	100	400	125	480
5	100	500	130	610
6	100	600	110	720
Total	600	600	720	720

Table 2.3 Forecast and Actual Sales (bias)

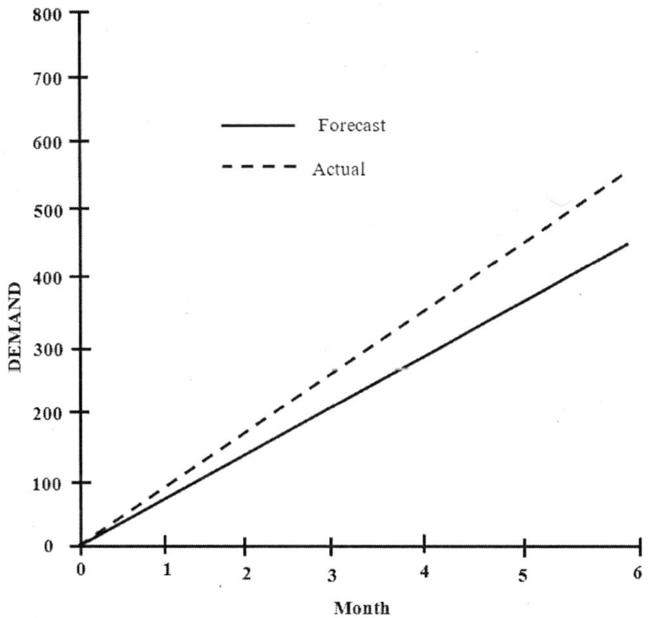

Figure 2.7 Forecast and actual demand with bias

When the real demand is consistently higher or lower than the predicted demand, this is called bias. If there is bias in the prediction, it should be looked at and maybe changed to make it more accurate.

By keeping an eye on the forecast, you can adjust your plans to account for errors and make the prediction more accurate. When a mistake or bias that is too big is seen, it should be looked into to find out what caused it.

There are often one-time, very special reasons why mistakes happen. Things like what a rival does, a customer shutting down, big one-time orders, and sales promotions are all examples. These reasons have to do with the talk about gathering and organizing data as well as the need to keep track of the events that led to the data. When this happens, the demand past needs to be changed to account for the unique situation.

Timing can also be a cause of errors. For instance, a winter that starts earlier or ends later will impact the timing of when people need snow shovels, even though the overall demand will remain unchanged. Monitoring the overall demand over time will help identify any mistakes in timing or any unusual occurrences that happened only once. Here is an example that demonstrates this. It should be noted that in April, the total demand returns to a typical range.

2.13.2 Random variation

In a given period, actual demand will vary about the average demand. The variation will depend on how often the product is bought. Demand for some goods will stay the same, and there won't be a lot of change. Some will be unsteady and have a lot of changes.

Look at Table 2.4, which shows the predicted and real demand. You can see that there is a lot of random change, but the error on average is 0. This proves that the average prediction was right and there was no bias. Figure 2.8 shows the graph of the data.

2.13.3 Mean Absolute Deviation

Forecast error needs to be measured before it can be used to make changes to the estimate or help with making plans. You can measure mistake in a number of different ways. Mean absolute deviation (MAD) is a method that is often used because it is easy to figure out.

Month	Forecast	Actual	Variation (Error)
1	100	105	5
2	100	94	-6
3	100	98	-2
4	100	104	4
5	100	103	3
6	100	96	-4
Total	600	600	0

Table 2.4 Forecast and Actual Sales Without Bias

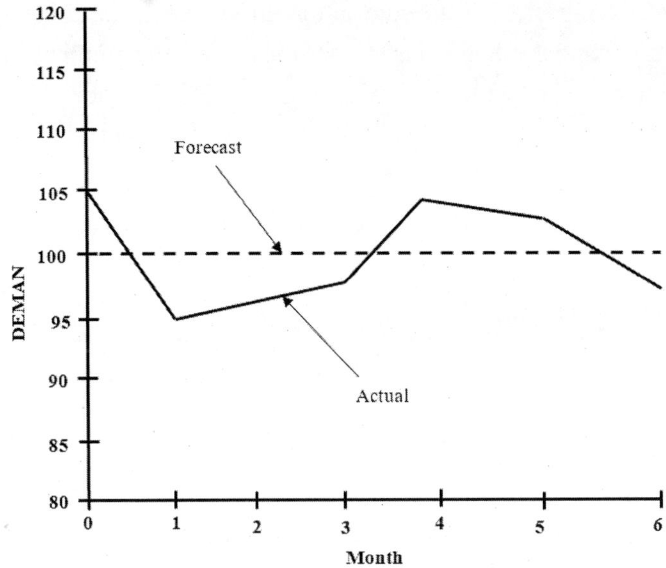

Figure 2.8 Forecast and Actual Sales without Bias

Look at table 2.4 for information on variations. There is still a lot of difference each month, even though the total error (variation) is zero. To find the difference, total error would not be useful. To find the variability, you can figure out the total mistake without taking into account the plus and minus signs and then take the average. The mean absolute variation is this:

- mean implies an average,
- absolute means without reference to plus and minus,
- deviation refers to the error:

$$\text{MAD} = \frac{\textbf{sum of absolute deviations}}{\textbf{number of observation}}$$

Example 2.7: Given the data shown in Figure 2.10, calculate the mean absolute deviation.

Answer: Sum of absolute deviations 5+ 6+2 +4+3+4=24

$$\text{MAD} = \frac{24}{6} = 4$$

Observe that the outcome differs significantly if the deviations are not interpreted as absolute numbers: Sum of deviations 5+ (- 6) + (- 2) + 4 + 3 + (- 4) = 0

This clearly shows that real deviations are not a good way to find the average forecast error, since they would show that the "average" forecast error is zero in this case. That's better done with the MAD. The fact that the sum of the differences is zero proves that there is no bias. In general, any sum of deviations over time that is not zero shows that there is prediction bias. When you subtract the forecast from the actual demand for a given time, you get the deviations. If the sum of the deviations is positive, it means that the actual demands are higher than the forecasts, which means that the method used to make the forecasts is biased toward the low side. If the sum of the differences is negative, on the other hand, it means that the method for making predictions is too optimistic. When combined with the MAD for the data, this kind of information can help a lot with planning for safety inventory.

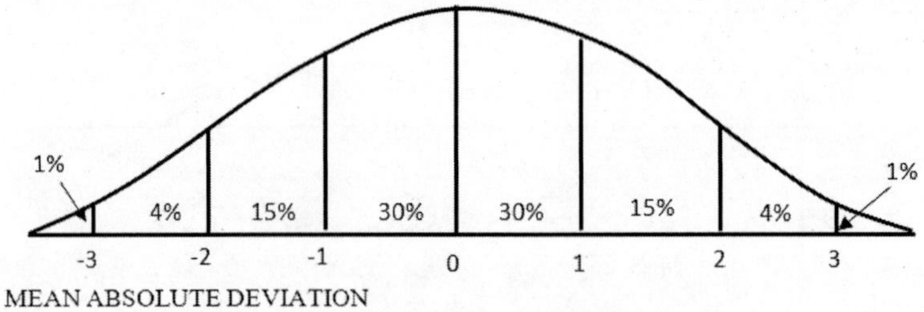

Figure 2.9 Normal Distribution Curve

2.14 Normal distribution

The mean absolute deviation shows how far off the prediction was from what actually happened. The actual desire isn't always close to what was predicted, but it is most of the time. A bell-shaped curve is made when you plot the number of times actual demand is a certain amount. Figure 2.12 shows what this kind of distribution is called a smooth distribution.

There are two important things to know about normal curves: the average or central trend and the spread or dispersion of the distribution. The prediction can be seen in Figure 2.12. It is the central trend. The standard deviation tells us how spread out the normal curve is, or how fat or thin it is. The standard deviation gets bigger as the spread gets bigger. It is easy to figure out and use the mean absolute deviation, which is a close estimate of the standard deviation.

From statistics it has been proven that the error will be within

- +1 MAD of the average about 60% of the time;
- +2 MAD of the average about 90% of the time;
- ±3 MAD of the average about 98% of the time.

2.15 Contingency planning

Assume that a prediction has been made stating that the demand for door slammers will be 100 units, while the capacity for producing them is 110 units. The historical calculation of the mean absolute divergence between the actual demand and the prediction has yielded a value of 10 units. This indicates a 60% probability that the actual demand will fall within the range of 90 to 110 units, while there is a 40% probability that it will not. Based on this information, manufacturing management might potentially develop a contingency plan to handle the potential increase in demand.

2.16 P/D Ratio

Due to the inherent margin of error in projections, companies who depend on them may encounter a range of issues. For instance, if the incorrect material is purchased, it could potentially be transformed into the incorrect products during the manufacturing process. The utilization of the P/D ratio is a more dependable method for generating the desired outcome.

P, or **production lead time**, is the stacked lead time for a product. It includes time for purchasing and arrival of raw materials, manufacturing, assembly, delivery, and some- times the design of the product. Figure 2.10 shows various times in different types of industries.

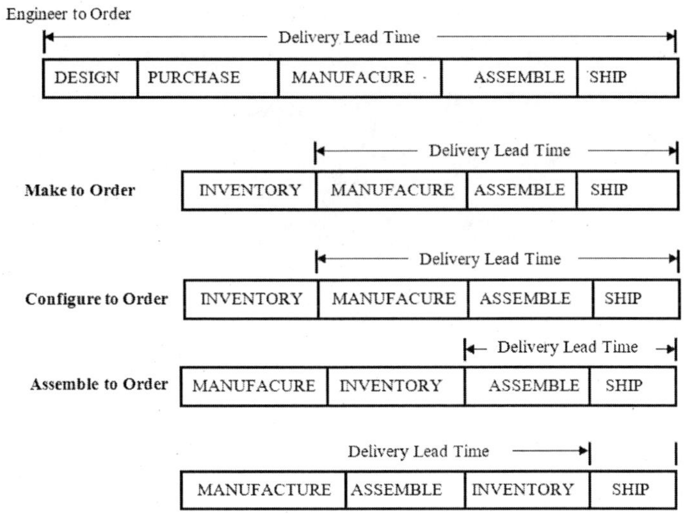

Figure 2.10 Manufacturing Lead Time and Strategy

D, also known as demand lead time, refers to the amount of time it takes for the client to receive their order. The lead time refers to the duration between when a customer initiates an order and when the items are delivered. The duration can vary significantly, ranging from brief periods in a make-to-stock setting to extended durations in an engineer-to-order organization

Adding safety stock to inventory has long been thought to be the best way to protect against the inherent mistake in forecasting. There is an extra cost for keeping extra stock "just in case." You could also try to be more correct with your predictions. You can go this way five different ways.

1. **Reduce P time.** As the duration of the P time increases, the probability of error also increases. Preferably, P should be smaller than D.
2. **Force a match between P and D.** Moving in this direction can be done in two ways:
   - Ensure that the customer's delivery timing matches your production time. This is a frequent occurrence with customized items, where the producer produces the product based on the exact requirements provided by the customer.
   - Offer for sale the items or products that you have predicted or projected. This will occur during the process of market regulation.

   An excellent illustration is the automotive industry. Manufacturers often provide incentives at the end of the automobile year to boost sales as anticipated.
3. **Simplify the product line.** There is more space for error in a product line with greater variety.

4. **Standardize products and processes.** Consequently, customization needs to take place in proximity to the completion of assembly. This method is alternatively referred to as delay. All components have same or comparable basic elements.
5. **Forecast more accurately.** Make forecasts using a well-thought-out, well-controlled process.

If P < D, then the product can be manufactured in a shorter duration than the anticipated customer lead time.

QR For Additional Learning on Advance Production Forecasting Techniques

2.17 Master Scheduling

The next step in the industrial planning and control process is to make a master production schedule (MPS). This comes after planning production. This part talks about some basic things you should think about when making and running an MPS. It is a very important tool for planning and allows sales and producers to talk to each other. The MPS is an important part of the method for planning production.

In other words, it connects what is planned for output to what is actually built. From this point of view, it's the main link between what customers want and what the factory can make.

- It's used to figure out how much space and money are needed.
- The material needs plan is based on it. The MPS and bills of material show what parts need to be bought and manufactured based on a list of things that need to be made.
- Priorities are kept up to date. The MPS is an important plan for making things.

The MPS works with finished goods, while the production plan deals with groups of products. In each family, it breaks down the production plan into the needs for each end item, along with the date and amount needed. The production plan sets limits on the MPS. This means that the number of things in the MPS should match the number shown on the production plan. By way of example, if the production plan says that 1000 tricycles will be made in a certain week, then the MPS should plan for 1000 individual types. Its goal is to find a balance between what the market wants (priorities) and what materials, labor, and tools are available (capacity) for making things.

The company puts together finished goods from bits that are made up of smaller parts. To keep up with the master production schedule, these must be ready at the right time and in the right amount. The schedule for these parts is made by the material requirements planning system based on what the MPS needs. In this way, the material needs plan is driven by the MPS.

The MPS is a plan for manufacturing. It shows what the market wants and what manufacturing can do. It also makes a priority list for manufacturing to follow. The MPS forms a vital link between sales and production:

- It enables the ability to provide accurate order commitments. The MPS is a comprehensive outline detailing the specific items to be manufactured and the corresponding schedule for production. Therefore, it provides information to sales and manufacturing regarding the anticipated availability of goods for delivery.
- It is a mutually agreed-upon arrangement and contractual agreement between the marketing and manufacturing departments.
- The MPS serves as a foundation for sales and production departments to ascertain what products are to be made. It is not intended to be inflexible. This is a communication device that serves as a foundation for implementing changes that align with market expectations and manufacturing capabilities. The necessary data for the development of a Master Production Schedule (MPS) is supplied by:
- The production plan refers to the aggregated production plan that is formulated during the Sales and Operations Planning (S&OP) process.
- Predictions for specific final products.
- Received orders from customers and for restocking inventory.
- Stock levels for specific finished products.
- Limitations on capacity.

2.18 Relationship to Production Plan

Assume that the following production plan is created for a group of three items for a family.

Week	1	2	3	4	5	6
Aggregate Forecast (units)	160	160	160	160	215	250
Production Plan	205	205	205	205	205	205
Aggregate Inventory (units)	545	590	680	680	670	625

Opening inventories (units) are

Product A	**350**
Product B	**100**
Product C	**50**
Total	**500**

The next step is to anticipate how much each item in the family will sell for. Remember that the forecast data for the production plan came from combining data from different product families. For the master schedule, you need a more detailed prediction for each item in the family. For the sales and operations planning process, estimates were made for groups of products, since predictions for groups of products are often more accurate than predictions for single products. For the S & OP, only product family forecasts were needed because it was only being used to plan the resources needed to make the product family prediction and not to make individual items. Also, keep in mind that the estimates for each item may not add up to the exact forecast for the whole family of products. This is because the production plan for the family limits the total number of items that can be made.

Week	1	2	3	4	5	6
Product A	70	70	70	70	70	80
Product B	40	40	40	40	95	120
Product C	50	50	50	50	50	50
Total planned	160	160	160	160	215	250

With this information, the master planner needs to make a plan that works with the restrictions. The picture below shows one possible option.

Master Schedule

Week	1	2	3	4	5	6
Product A						205
Product B	205	205	205			
Product C				205	205	
Total planned	205	205	205	205	205	205

Inventory

Week	1	2	3	4	5	6
Product A	280	210	140	70	0	125
Product B	265	430	595	555	460	340
Product C	0	-50	-100	55	210	160
Total planned	205	205	205	205	205	205

This schedule is satisfactory for the following reasons:

- It regulates the plant's production of particular items by determining the appropriate times to begin and end production.
- The capacity aligns with the production plan.
- It is inadequate due to the following reasons:
- The inventory balance is worse in comparison to the entire inventory.
- Product C experiences stockouts during periods 2 and 3.

The term "master production schedule" refers to the final row of the matrix. Master scheduling is the procedure of determining the final production schedule. Hence, the comprehensive matrix is referred regarded as a master schedule.

2.19 Developing A Master Production Schedule

The aims in creating an MPS are as follows:

- To uphold the intended standard of customer service, it is important to either manage the inventory of finished goods or create a schedule that aligns with client delivery demands.
- In order to optimize the utilization of resources such as materials, labor, and equipment.
- In order to sustain the necessary levels of inventory investment.

In order to achieve these goals, the plan must meet consumer demand, adhere to manufacturing capacity, and comply with the production plan criteria. The process of preparing an MPS involves three distinct steps:

- Create an initial Master Production Schedule (MPS).
- Verify the initial Master Production Schedule (MPS) in comparison to the existing capacity.
- Address discrepancies between the initial Master Production Schedule (MPS) and the available production capability.

2.19.1 Preliminary Master Production Schedule

An illustrative example is employed to demonstrate the process of generating a Master Production Schedule (MPS) for a product that follows a make-to-stock approach, involves inventory management, and is produced in batches.

A particular item is made in groups of 100, and 80 units are supposed to be in stock when the store opens. Figure 3.1 shows the expected demand, the expected amount on hand, and the estimated MPS.

Period I start with 80 units in stock. Once the expected 60 units that are needed are sold, 20 units are expected to be available. Another predicted demand of 60 in period 2 is not met, so an MPS receipt of 100 for week 2 needs to be planned. This means that 60 units should be available at the end of time 2 (20+100-60 = 60). In time 3, the expected demand for 60 is met by the anticipated 60 that is already on hand, leaving an anticipated available of 0. In period 4, another 100 must be received. Once the expected demand of 60 units is met, 40 units will still be in stock.

This process of making an MPS is done for every member of the family. If the total amount of items planned to be made and the total amount of items left in stock do not match. On hand 80 units

Lot size = 100 units

Period		1	2	3	4	5	6
Forecast		60	60	60	60	60	60
Project Available	80	20	60	0	40	80	20
MPS			100		100	100	

Table 2.5 MPS Example

If everyone agrees with the production plan, some changes need to be made to each plan so that the total production stays the same.

After the first drafts of the master production plans are made, they need to be compared to the capacity that is available. A rough-cut capacity planning is what this method is called.

Example 2.8: A bunch of different kinds of nut crackers are made by Amalgamated Nut Crackers, Inc. The walnut type is the most popular, and the sales team has made a 6-week forecast. The inventory at the start is 50 dozen, which is the measure used for planning. You have to make an MPS as the master planner. There are 100 dozen of the nutcrackers in a lot.

Answer

Period		1	2	3	4	5	6
Forecast Sales		75	50	30	40	70	20
Project Available	50	75	25	95	55	85	65
MPS		100		100		100	

2.20 Production Planning, Master Scheduling and Sales

The production plan aligns the overall predicted demand with the existing resources. The overall plan for production is derived from the strategic plan, strategic business strategy, and market projections, incorporating the necessary information to determine what products will be manufactured in order to fulfill the projected demand. It is a significant result of the sales and operations planning process. The fulfillment of the predicted demand is contingent upon the forecast itself and must be strategically planned within the constraints of available resources. It does not focus on the specific details of what will be produced. The purpose of this is to establish a structure that allows for the creation of specific plans inside the MPS.

The MPS is constructed using projections and real demands for specific end products. It aligns demand with the production plan and available resources to create a manufacturing strategy that can be successfully executed. The MPS focuses on determining the specific items to be produced, their quantities, and the timing of production in order to fulfill projected demand.

The production plan and the Master Production Schedule (MPS) separate the sales forecast from manufacturing by creating a manufacturing plan. Collectively, they strive to harmonize the existing resources of machinery, infrastructure, workforce, and materials with the projected level of demand. However, they do not serve as a prediction of sales, nor do they necessarily align with what is sought. The MPS is a strategic blueprint outlining the intended actions and capabilities of the production department. Figure 2.11 shows the relationship among the sales forecast, production plan, and MPS.

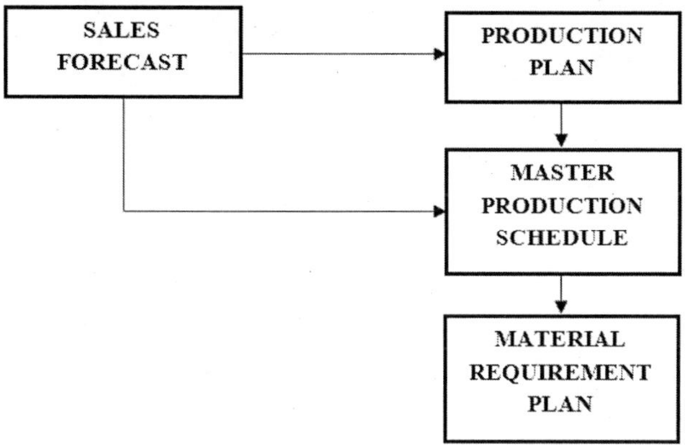

Figure 2.11 Sales Forecast, Production Plan, and Master Production Scheduling

The manufacturing potential and willingness must be taken into account by the MPS. If it isn't, there will be overflowing capacity plans, missed deadlines, erratic delivery assurances, spikes in shipments, and a dearth of accountability. The MPS is a plan for particular end products or "buildable" parts that manufacturing anticipates producing in the future. This is the time when marketing and manufacturing have to decide which final products will be produced. Producing goods is the focus of manufacturing, and selling goods is the goal of marketing. Still, the MPS is not designed to be inflexible. Demand fluctuates, production issues arise, and occasionally parts are in short supply. The MPS may need to be changed as a result of these occurrences. Sales and production must fully comprehend and approve any changes that are made. The MPS offers a plan that everyone can agree upon and a foundation for making changes.

2.21 Production Activity Control

Following the completion of all the necessary planning and scheduling, it is now the appropriate moment to implement the plans. Production activity control (PAC) is tasked with carrying out the master production schedule (MPS) and the material needs plan (MRP). At the same time, it must make good use of people and machines, minimize work-in-process inventory, and maintain customer service.

- The material requirements plan authorizes PAC:
- To initiate the issuance of work orders to the workshop for the purpose of manufacturing.

- To oversee work orders and ensure their timely completion.
- The role entails assuming responsibility for the meticulous and prompt planning of order flow within the production process, executing the plan, and overseeing the job until its successful conclusion.
- To oversee daily operations and offer essential assistance.

Figure 2.12 shows the relationship between the planning system and PAC. The planning, implementation, and control functions can be used to categorize the PAC system's operations.

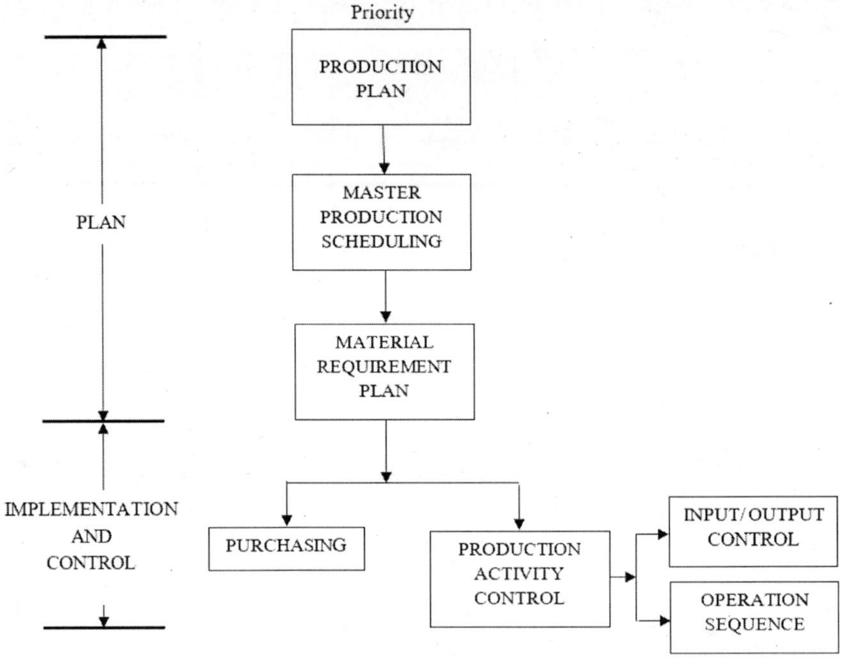

Figure 2.12 Priority and Production Activity Control

2.21.1 Planning

The flow of work through each of the work centers must be organized to match delivery deadlines, which means production activity control must conduct the following:

- Ensure that the relevant materials, tooling, staff, and information are available to manufacture the components as needed.
- Schedule start and completion dates for each shop order at each work center so the scheduled completion date of the order can be fulfilled. The planner will be responsible for creating a load profile for the work centers.

2.21.2 Implementation

After the plans are finalized, production activity control is responsible for implementing them by instructing the shop floor on the necessary tasks. Instructions can be conveyed either through a shop order including the necessary details or by generating a schedule that presents information about the product, quantities, and dates.

- Production activity control will collect the necessary information required by the shop floor to manufacture the product.
- Authorize the release of orders to the shop floor in accordance with the material requirements plan. This process is referred to as dispatching.

2.21.3 Control

After the designs are finalized and the shop orders are issued, it is essential to closely monitor the process to gain accurate insights into the actual progress. The results are evaluated against the plan to see if any remedial measures are required. Production activity control will encompass the following tasks:

- Arrange the shop orders in the preferred order of importance according to the work center and create a dispatch list using this information.
- Monitor the real-time execution of work orders and analyze it against the predetermined timetables. If it is deemed necessary, PAC must implement corrective measures such as revising plans, amending schedules, or modifying capacity in order to fulfill final delivery obligations.
- Supervise and manage the progress of unfinished tasks, the time it takes to complete them, and the waiting lines at work stations.
- Provide a report on the efficiency of the work center, including operation times, order amounts, and scrap.

The functions of planning, implementing, and controlling are shown in Figure 2.13

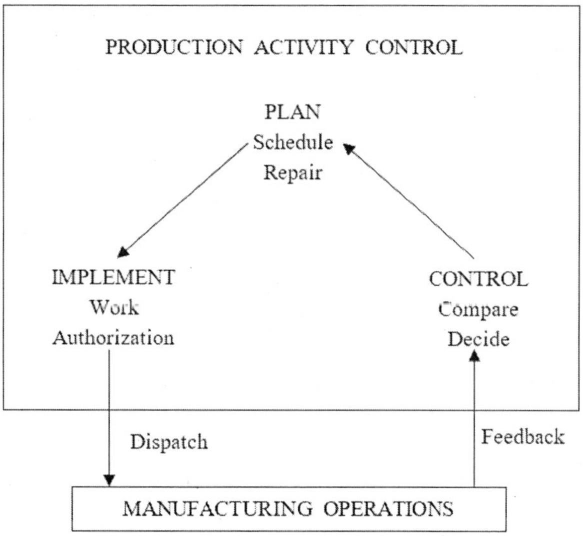

Figure 2.13 Schematic diagram of a Production Control System

2.22 Manufacturing Systems

Each company uses a different type of production control system, but they should all be able to do the things we just talked about. How important these functions are, though, will rely on the type of manufacturing process. The steps used in manufacturing can be easily grouped into three groups:

a) Flow manufacturing.
b) Intermittent manufacturing.
c) Project manufacturing.

2.22.1 Flow manufacturing

Flow manufacturing is the process of making a lot of standard goods at once. The process is called repetitive manufacturing when the units are separate, like cars and tools. It is called continuous manufacturing when the units are made all the time, like gasoline. The four main things that make up flow production are:

- Routing is set in the prescribed area, and work areas are set up based on the routing. Working at one work center takes almost the same amount of time as working at any other center in the line. This keeps the flow going smoothly.
- Work centers are places where only a few similar goods are made. The machinery and tools are made to make those specific goods.
- Some kind of mechanical transfer moves things from one desk to another. Stock of work-in-progress doesn't build up too quickly, and processing times are low. 4. The line sets the capacity.

Controlling production activity means planning the flow of work and making sure that the right materials are sent to the line at the right time. It's pretty easy to set up and manage because work instantly moves from one workstation to another.

2.22.2 Intermittent manufacturing

There are a lot of changes in product design, process needs, and order quantities that happen during intermittent manufacturing. The following things describe this type of manufacturing:

- The shop has a wide range of work flows that depend on the design of each item. There may be times when orders take longer to handle at one workstation than at another. This means that the flow of work isn't even.
- The machines and people who work with them need to be able to adapt to the different tasks that come up with intermittent production. Usually, machines and work centers are put together by what they do. For example, all lathes are in the same area.
- Most of the time, throughput times are long. It's hard to schedule work to come just when it's needed because each work center takes a different amount of time on each order and work often queues up before it gets to the center, which causes long delays in processing. A lot of work-in-process material is kept on hand.

- The amount of space needed relies on the mix of products being made, and it can be hard to guess what that mix will be.

It's hard to keep track of production activities in irregular manufacturing. PAC is a big part of this type of manufacturing because of the large number of goods made, the different routes, and the problems with scheduling. Shop orders or precise schedules for each batch being made are often used for planning and controlling. This kind of setting is assumed for most of the discussion of PAC in this book.

2.22.3 Project manufacturing

Project manufacturing often entails the production of a single unit or a limited quantity of units. An instance of intricate ship construction is provided. Close collaboration between production, marketing, purchasing, and engineering is necessary due to the iterative nature of product design.

Project manufacturing or management employs numerous strategies similar to production activity control, but also possessing distinct characteristics. Common activities encompassed in project management often include:

- Commencing the project by determining the project prerequisites • Formulating the project strategy, encompassing the project's extent, timetable and assignments, financial plan, resources, and potential hazards
- Implementing the project by performing the necessary tasks
- Overseeing and managing the project activities and resources, and effectively conveying the project's progress to stakeholders
- Concluding the project by recording the outcomes, as well as any deviations in time and expenses

2.23 Data Requirements

In order to strategize the processing of materials during the production process, PAC requires the following information:

- Determining the goods or services to be produced and the quantity to be produced.
- When components are required to ensure the deadline is achieved.
- What are the necessary procedures for manufacturing the product and what is the estimated duration of each procedure?
- What are the available capabilities of the different labor centers?

An operational control system is essential for production activity management. Typically, the information required to address these inquiries is structured and stored in databases. The databases contain two types of information: planning and control.

2.24 Planning Information

Four essential types of planning information required include: item master, product structure, routing, and work center master.

2.24.1 Item master

Each part number in the item master database corresponds to a single record that includes all relevant data about the part. For the Political Action Committee (PAC), the following items are included:

- Part number. A component is designated a unique number.
- Description of the component.
- Lead time for manufacturing. The typical time required to fabricate this component.
- Quantity of lots. The quantity that is typically purchased in a single transaction.

2.24.2 Product structure (bill of materials)

The bill of material (product structure) contains an inventory of the single-level components and the quantities required to assemble a parent. It serves as the foundation for a select list that storeroom personnel will employ to gather the necessary components for the assembly. A bill of material can be presented in a variety of forms, such as a single level bill of material, an indented bill of material, transient bill of material, matrix bill of material, and costed bill of material. The bill of material is referred to as a formula, recipe, or ingredients inventory in certain industries, particularly the process industry.

2.24.3 Routing

Each part manufactured is represented by a record in the routing. The routing comprises a sequence of procedures necessary to manufacture the item. Each product is accompanied by a detailed set of instructions that outline the step-by-step process of its production. The information provided includes the following details:

The manufacturing process involves the necessary processes and their specific sequence.

- A concise overview of each operation.
- The necessary equipment, tools, and accessories required for each specific task.
- Setup times refer to the normal duration needed to prepare the equipment for each operation.
- Run times refer to the normal duration needed to complete one unit of processing for each operation. Duration for each individual surgery.

2.24.4 Work center master

The work center master compiles all pertinent information for a work center. It provides specific information for each work center about the following:

- Identification number of the work center.
- Maximum volume or amount that can be held or accommodated.
- Weekly shift count.
- Quantity of machine hours worked throughout each shift.
- Labor hours per shift.
- Optimization.

- Usage.
- Queue time refers to the average duration that a job has to wait at the work center before work is initiated.
- Alternate work centers refer to work centers that can be used as substitutes or options.

2.25 Control Information

Control in intermittent manufacturing is exercised through shop orders and the data contained on these orders.

2.25.1 Shop order master

Every manufacturing order that is currently in progress has a corresponding entry in the shop order master. The objective is to furnish condensed data regarding each shop order, encompassing the following details:

- The shop order number is a distinct identifier assigned to each shop order.
- Quantity of items to be ordered.
- Task finished.
- Amount discarded.
- Amount of material allocated to fulfill the order.
- The due date refers to the anticipated completion date of an order.
- Priority is a numerical value that is assigned to determine the relative order or importance of something in comparison to others.
- Outstanding amount remaining for the unfinished quantity.
- Information on costs.

2.26 Order Preparation

After receiving authority to handle an order, production activity control is responsible for planning and arranging its release to the shop floor. The order should undergo a thorough evaluation to ensure that the required tooling, material, and capacity are readily accessible. If the conditions are not met, the order cannot be fulfilled and should not be authorized for release.

By utilizing MRP software, the availability of material is automatically verified and assigned to a shop order, eliminating the need for any additional verification. In the absence of MRP software, production activity control is required to manually verify the availability of all the components required for the manufacturing of goods.

If a capacity requirements planning system has been utilized, the required capacity should be accessible. However, at this time, there may be discrepancies between the anticipated capacity and the actual availability, which can be attributed to delayed product delivery, daily fluctuations in workforce, and other factors. In the absence of capacity requirements planning, it is imperative to ascertain the availability of capacity prior to releasing orders.

Verifying capacity availability is a two-step procedure. Initially, it is necessary to arrange the order in a predetermined timeframe to determine the required capacity. Subsequently, the workload on work centers must be assessed throughout that specific period.

2.27 Objectives of Scheduling

The primary goal of scheduling is to ensure timely delivery and optimize the utilization of industrial resources. It entails determining the commencement and conclusion dates for every operation necessary to finalize a task. In order to create a dependable timetable, the planner needs to possess data regarding the route, necessary and accessible capacity, competing tasks, and manufacturing lead times at each work center involved.

2.27.1 Manufacturing Lead Time

Manufacturing lead time is the amount of time it usually takes to make a standard lot of an item. Most of the time, it has five parts:

1. Queue time is the amount of time a job has to wait at a work place before it can be done.
2. Setup time is the amount of time needed to get the work area ready to use.
3. Run time is how long it takes to carry out the action on the order.
4. The wait time is how long the job stays at the work center before it is moved to the next work center.
5. Time to move or get from one work area to another.

The overall manufacturing lead time will be calculated by adding the time required for order preparation and release to the manufacturing lead time for each individual activity.

The element with the greatest magnitude among the five is queue time. In an intermittent manufacturing operation, the intermittent process usually makes up 85-95% of the entire lead time. Production activity control is the department in charge of overseeing the queue management process by regulating the inflow and outflow of work at work centers. Reducing the amount of orders in the queue (load) also reduces the queue time, lead time, and work-in-process. Expanding capacity also decreases waiting time. Production activity control is responsible for managing the intake of orders into the production process and overseeing the available capacity to regulate the queue and work-in-progress.

Example 2.9

On work centers A and B, a request is made for 100 units of a product. For A, the setup takes thirty minutes, and each unit needs ten minutes to operate. B takes fifty minutes to set up and five minutes to complete each piece. There is a 4-hour gap between the two procedures. Ten minutes is how long it takes to get from point A to point B. The wait period after operation B is four hours, while the time needed to move the merchandise into stores is fifteen minutes. At neither workstation is there a queue. Calculate the total time required to complete the order's manufacturing.

Answer

- Work center A process time = 30+ (100 x 10) = 1030 minutes
- Hold time = 240 minutes
- Move time from A to B = 10 minutes
- Work center B process time = 50+ (100 x 5) = 550 minutes
- Hold time = 240 minutes
- Shift time from B to stores = 15 minutes
- Total production lead time = 2085 minutes

  = 34 hours, 45 minutes

2.28 Scheduling Techniques

Various methodologies are employed to arrange shop orders inside a manufacturing facility, all of which necessitate comprehension of forward and backward scheduling, as well as limited and infinite loading.

2.28.1 Forward scheduling

It is based on the assumption that the procurement of materials and the scheduling of operations for a component begin as soon as the order is received, regardless of the due date. Additionally, operations are scheduled to occur in a forward direction from this starting date. The procedure is illustrated by the first line in Figure 2.14. Completing the task ahead of schedule often leads to an accumulation of inventory. This method is employed to determine the earliest possible date of delivery for a product. Forward scheduling is employed to determine the duration required to finish an activity. This technique is employed to determine promise dates for customers or assess the feasibility of catching up with orders that are behind schedule.

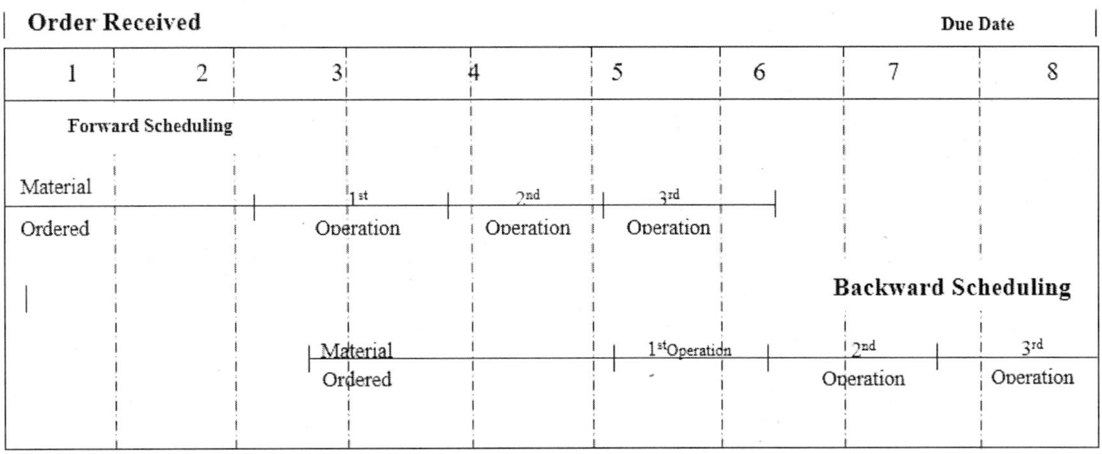

Figure 2.14 Forward and Backward Scheduling: Infinite Loading

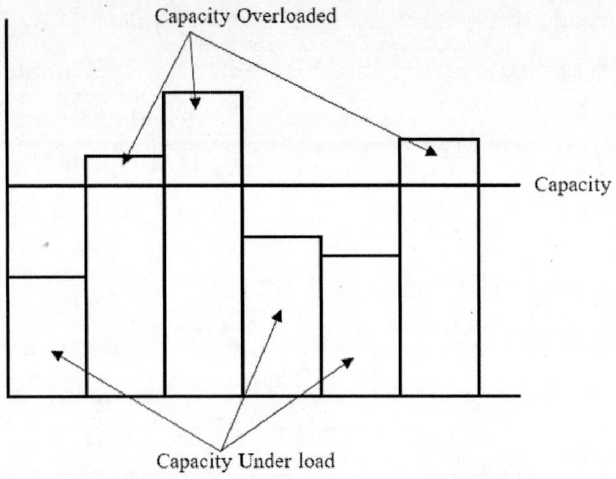

Figure 2.14 Infinite Load Profile

2.28.2 Backward scheduling

The second line in Figure 2.14 demonstrates the concept of backward scheduling. The final action in the routing is prioritized and is planned to be finished by the specified deadline. Operations prior to the last operation are scheduled in reverse chronological order. This system employs the same logic as the Material Requirements Planning (MRP) system and arranges goods to be accessible when required. The reduction of work-in-process inventory may lead to a potential decline in customer service due to the limited availability of spare time in the system. Backward scheduling is employed to ascertain the precise commencement time for an order. Backward scheduling is prevalent in a make-to-stock setting because to its ability to minimize inventory levels.

Figure 2.15 also demonstrates the concept of infinite loading. It is assumed that the workstations where operations 1, 2, and 3 are carried out have sufficient capacity when needed. It fails to take into account the presence of other shop orders vying for capacity at these work locations. It presupposes that there would be an unlimited amount of capacity accessible. Figure 2.15 displays a load profile for a capacity that is unlimited. Observe the excessive and insufficient load.

Finite loading operates under the assumption that there is a specific and fixed limit to the capacity that is available at each workstation. If the workstation does not have sufficient capacity due to other shop orders, the order must be rescheduled to a different time period. Figure 2.16 depicts the state of affairs. In the given example of forward scheduling depicted in Figure 2.16, the initial and subsequent activities cannot be executed at their designated workstations as planned due to the unavailability of the necessary capacity at the scheduled time. These operations need to be rescheduled for a later time frame. Similarly, in the case of back scheduling, the second and first operations are unable to be executed at their designated times and must be rescheduled to an earlier time period. Figure 2.17 displays a load profile for finite loading. Observe that the load is evenly distributed, hence preventing any instances of overload.

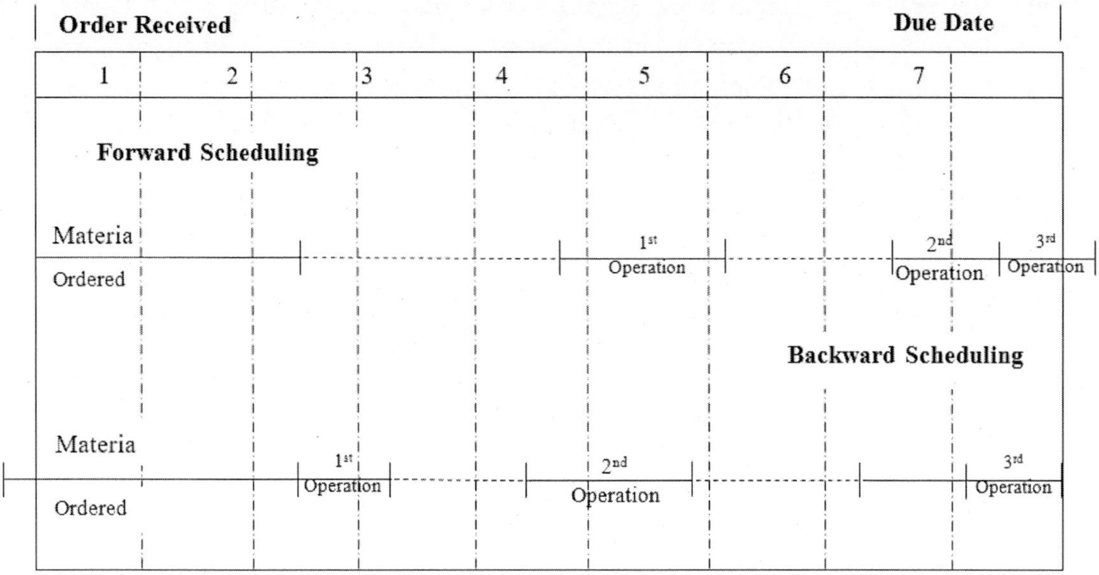

Figure 2.15 Forward and backward scheduling

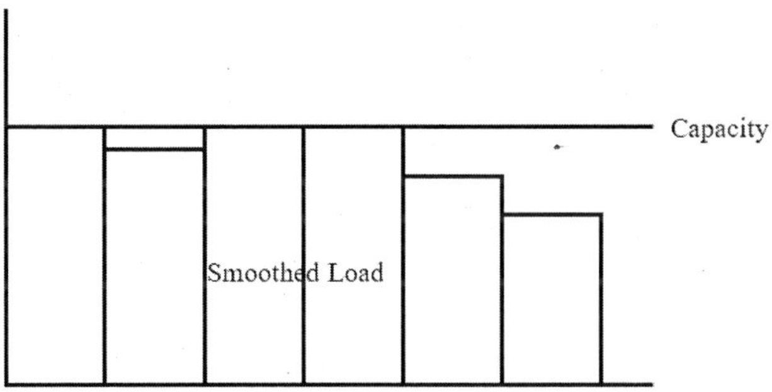

Figure 2.16 Finite Load Profile

Example 2.10: The company has received an order for 50 units of brand X, which is scheduled to be delivered on day 100. Create a reverse chronological timeline using the given information:

a. Each operation is given to one machine.
b. The factory operates on a single 8-hour shift for 5 days each week.
c. The components are transported together in a batch of 50.

Answer

Part	Operation	Time (days)
A	10	5
A	20	3
B	10	10
X	Assembly	5

Every industry faces the challenges of resource scheduling, balancing demand and supply, and managing available vs necessary capacity. In the transportation business, efficient scheduling and routing of truck fleets is crucial to optimize costs, ensure timely deliveries, minimize downtime, and reduce unproductive time associated with empty vehicle returns.

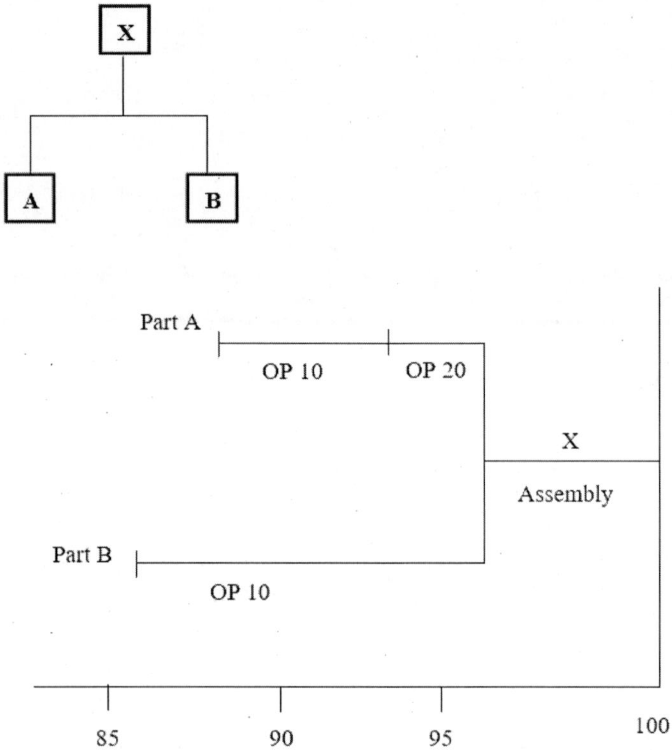

Figure 2.17 Scheduling in a Nonmanufacturing Setting

In the health care business, organizations must effectively manage the allocation of resources such as doctors, surgeons, nurses, technicians, operating rooms, and hospital rooms. This involves ensuring that the capacity of these resources aligns with the changing demands of patients, emergency vehicles, and significant traumas. Although certain events, like arranging office visits and annual physicals, can be anticipated, a significant portion of the workload arises from unforeseen circumstances such as diseases and natural disasters, making it challenging to predict. Several hospitals have started predicting the workload by analyzing the historical data of patient days every month to identify patterns or seasonal variations, with the aim of improving capacity planning. Time studies can also be conducted to establish benchmarks for activities such as laboratory work and surgical preparation, in order to accurately assess the capacity of certain resources. Effective resource planning is crucial in-service businesses such as retail, food, aviation, and similar sectors. Service professionals are typically scheduled on a weekly, daily, and hourly basis, taking into account projections of when consumers are most likely to require the service. In this instance, the capacity component mostly consists of human resources, although it may also encompass equipment, tools, and time. Certain industries, such as aviation and transportation, must adhere to restrictions on personnel working hours, such as a maximum limit on the number of hours worked in a day.

A corporation can maximize resources and boost capacity in a service industry by adapting the level of workforce through cross-training personnel, deploying part-time workers, or implementing automation solutions such as self-help kiosks. Personnel can be utilized by performing nonurgent duties, such as cleaning and maintenance, during periods of low or no demand. A well-known food chain devised its own scheduling system, which encompassed breaks, forecasts for food preparation, and determining the optimal time to reduce the production of baked products and start providing samples to clients.

2.29 Scheduling Bottlenecks

In intermittent production, achieving a balance between the capacities of different workstations and the demand for their capacity is extremely challenging. Consequently, certain workstations are experiencing excessive load while others are seeing insufficient load. Bottlenecks refer to the overcrowded workstations where the needed capacity exceeds the available capacity. A bottleneck refers to a facility, function, department, or resource that is unable to meet the demand placed upon it due to its limited capacity.

2.29.1 Throughput

Throughput refers to the overall quantity of goods or services that are produced and flow through a certain facility. Bottlenecks regulate the rate at which all goods are processed, as the whole throughput cannot exceed the capacity of the bottleneck. When work centers that supply bottlenecks generate more than the bottleneck can handle, there is an accumulation of surplus work-in-process inventory. Thus, it is advisable to arrange work according to the capacity of the bottleneck to ensure efficient processing. Work centers that are supplied by bottlenecks have their production rate limited by the bottleneck, and their schedules should be synchronized with the bottleneck's timetable.

Example 2.11: Assume that a producer manufactures wagons consisting of a box body, a handle assembly, and two-wheel assemblies. The weekly demand for wagons is 500 units. The weekly production capacity for wheel assembly is 1200 sets, for handle assembly is 450 sets, and for final assembly is 550 wagons.

a) What is the capacity of the factory?
b) What limits the throughput of the factory?
c) How many wheel assemblies should be made each week?
d) What is the utilization of the wheel assembly operation?
e) What happens if the wheel assembly utilization is increased to 100% ?

ANSWER

a) The weekly production is 450 units.
b) The handle assembly operation sets the upper limit on the throughput.

c) A total of 900 wheel assemblies should be manufactured on a weekly basis. This corresponds to the capability of the handle assembly operation.
d) The utilization of the wheel assembly procedure is 75%, calculated by dividing 900 by 1200.
e) Accumulation of surplus inventories occurs.

The service sector also include throughput, which refers to many factors such as the duration of a patient's stay at a hospital, the frequency of table turnover at a restaurant during peak dinner hours, or the waiting time for a customer in a bank queue. An obstacle faced by service businesses is the unpredictability in the duration of service delivery. The duration of customer waiting time at a bank, such as the time spent in line, can significantly differ based on factors such as the quantity and nature of transactions.

2.29.2 Bottleneck principles

Given that bottlenecks have a significant impact on the rate at which a facility can process tasks, it is crucial to consider certain fundamental principles:

1. The utilization of a non-bottleneck resource is not determined by its capacity, but rather by another constraint inside the process. In the preceding example problem, the utilization of the wheel assembly procedure was dictated by the handle assembly operation.
2. Utilizing a non-constrained resource continuously does not result in complete usage. If the wheel assembly procedure was operated continuously, it would produce 1200 sets of wheels every week, exceeding the required amount by 300 sets. Due to the accumulation of goods, this operation would inevitably need to cease.
3. The facility's capacity is determined by the capacity of the bottleneck. In the event of a malfunction in the handle assembly operation, the factory's production capacity is diminished.
4. Efficiency gained at a location that is not a bottleneck does not increase the overall capacity in other areas. Increasing the capacity of the wheel assembly operation to 1500 units per week by the industrial engineering department would not result in any gain as the excess capacity would remain unused.
5. Capacity and priority are interdependent factors that should be taken into account simultaneously. Assume that the wagon maker produced wagons with two distinct types of handles. During the setup process, no output is generated, hence diminishing the system's capacity. Due to the handle assembly being the limiting factor, each setup in this operation decreases the system's throughput. Optimally, the corporation would employ a single design of handle for a duration of six months, and subsequently transition to the alternative design. However, clients desiring the alternative kind of handle may not be inclined to endure a six-month waiting period. An optimal solution is required where the duration of runs is maximized while ensuring that priority (demand) is met.
6. Loads can and should be divided. Assuming that the handle assembly process, which is the limiting factor, manufactures one type of handle for a duration of two weeks before transitioning to the second type. The batch size consists of 900 handles. Instead of waiting for

all 900 units to be created before transferring them to the final assembly area, the manufacturer has the option to move a batch of 90 units per day. The process batch size (900) and the transfer batch size (90) are distinct. Consequently, the delivery of goods to the final assembly is synchronized with their utilization, resulting in a decrease in work-in-process inventory.

7. The primary emphasis should be on achieving equilibrium in the movement of customers within the shop. The crucial factor is the overall amount of output that ultimately results in marketable products.

2.29.3 Managing bottlenecks

Given the significant impact bottlenecks have on system performance, it is crucial to prioritize and manage them effectively. The following tasks must be completed:

1. Implement a period of time to allow for preparation or adjustment before encountering any bottleneck. A time buffer refers to an inventory or queue that is positioned before a bottleneck in a system. To ensure the continuous operation of the bottleneck, it is crucial to avoid any interruption in the flow of materials from the feeding workstations. The time buffer should be equal to the anticipated delay caused by feeding workstations, and no longer. The time buffer guarantees that the bottleneck will not cease operations due to insufficient work, and it maintains the queue at a predefined minimum level.
2. Regulate the flow rate of the material entering the bottleneck. In order to maintain a constant time buffer, it is necessary to supply a bottleneck with a rate of input that matches its capacity. The initial operation in the series of operations is referred to as a gateway operation. The operation in question, as well as any preceding activities, are responsible for controlling the flow of work into the bottleneck. It is crucial that these operations execute at the same rate as the bottleneck's output in order to maintain the time buffer queue.
3. Make every effort to ensure the necessary bottleneck capacity is provided. Any factor that enhances the capacity of the constraining element also enhances the capacity of the whole process. Enhanced utilization, reduced setups, and optimized techniques for minimizing setup and operational time are effective approaches for enhancing capacity.
4. Modify the weights. This is analogous to item 3, but it prioritizes the reduction of workload on a bottleneck by implementing strategies such as utilizing different work centers and subcontracting. Although utilizing non-bottlenecks may incur higher costs compared to using the bottleneck, it leads to better utilization and throughput of the entire facility. This, in turn, improves operational efficiency and has the potential to boost sales and profitability.
5. Modify the timetable. As previously mentioned, this should only be used as a last resort, but it may be essential to ensure precise delivery commitments.

After determining the available capacity of the bottleneck and the market demand it needs to meet, the scheduling of non-bottleneck resources can take place. Once a work order is finished at the bottleneck, it can be arranged for following processes.

Feeding operations must safeguard the time buffer by planning their schedule in reverse, starting from the bottleneck. If the time buffer is set to four days, the operation that occurs right before the bottleneck is scheduled to finish producing the necessary pieces four days prior to when they are planned to be processed at the bottleneck. Each previous operation can be rescheduled in the same manner so that the parts are ready when needed for the subsequent operation.

The feeding procedures can handle any disruptions without affecting the throughput, thanks to the time buffer. Additionally, there is a decrease in work-in-process inventories. By restricting the queue to the time buffer, lead times are decreased.

Bottlenecks are a common occurrence in all types of processes, including those seen in hospitals, banks, and restaurants. It is imperative to effectively manage them in order to maintain client loyalty. If the backlog is caused by insufficient manpower, it may be feasible to provide extra hours or implement additional shifts. Nevertheless, there are situations where expanding capacity, such as incorporating additional hotel rooms or airline seats, proves to be a challenging task.

A major airline identified a bottleneck in the aircraft turnaround process, which was restricting the number of possible flights and resulting in excessive downtime for their limited resource – flight crews. The entire process, starting from the moment the airplane arrived at the airport terminal, including passenger disembarkation, aircraft cleaning, caterer resupply, refueling, luggage unloading and reloading, and preparation for the next group of passengers, required a minimum of 45 minutes. It was established that there was a clear association between aircraft turnaround efficiency and schedule punctuality. By leveraging technology, optimizing the arrangement of luggage carts, and enhancing communication between flight crews and ramp personnel, they successfully reduced the average time to 20 minutes. The on-time departure percentage experienced a 30% improvement as a result of this.

2.30 Theory of Constraints

The segment on handling bottlenecks was formulated using the research conducted by Eliyahu M. Goldratt in his Theory of Constraints. It has enabled several individuals to reconsider their strategies for enhancing and overseeing their production procedures. The underlying principle of the work is that every action involved in creating a product or service consists of interconnected processes. Every process has a distinct capacity to provide the specified output for the operation, and in almost all cases, there is one process that restricts or limits the overall throughput of the activity. Refer to Figure 2.18 to observe an example of a process that produces product A.

The overall operation is limited by process 3, which has a capacity of four units per hour. Regardless of the level of efficiency in other processes and the number of process improvements implemented in processes 1, 2, and 4, it will always be impossible to surpass the total limit.

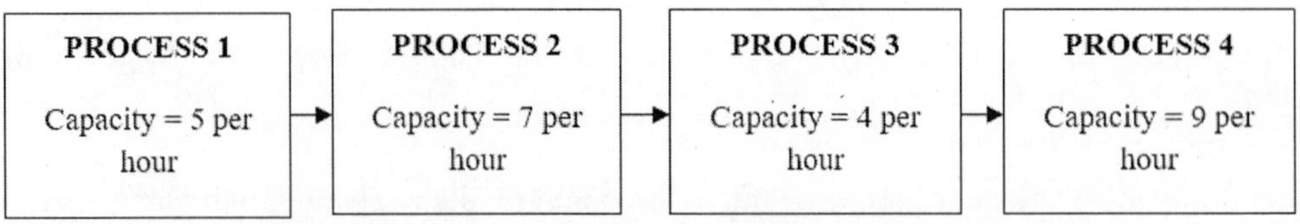

Figure 2.18 An operation producing product A

The operational output is four units per hour. Enhancing the efficiency and utilization of processes 1 and 2 will solely result in a rise in inventory levels, without any impact on sales.

Determining the limitation in a process can be quite straightforward. A final product always requires a predetermined sequence of steps, or processes, to be completed. The identification of a constraint occurs when a process is found to be operating at maximum capacity while inventory accumulates behind it, causing downstream operations to have idle time in processing the inventory. If all orders have been scheduled and all raw materials for those orders have been issued, but there is still idle time in all stages of a production sequence for the required production, then the constraint is said to be in the sales department.

2.30.1 Manage the Constraint

The section on Bottleneck Principles explored the development of some key criteria for effectively managing a limiting process or bottleneck. The principles of maintaining a balanced overall flow and ensuring consistent work at the constraint are crucial to the Theory of Constraints.

2.30.2 Improve the Process

After identifying a constraint, it is recommended to follow a five-step method to enhance the performance of the operation. The five steps can be succinctly summarized as:

1. Identify the constraint. An exhaustive analysis of the entire process is necessary to identify the specific process that restricts the rate of output. This notion extends the examination of this process beyond only the operational processes. For instance, in Figure 2.22, let's assume that the sales department was only able to sell the output at a rate of three units per hour. Therefore, sales would serve as the limiting factor rather than process 3. It is important to note that a restriction restricts the rate of output, rather than the amount of inventory or the level of production.

2. Exploit the constraint. Discover strategies to optimize the exploitation of the restriction in order to increase productive throughput. For instance, in numerous operations, all procedures are halted during the lunch break. If a process is identified as a constraint, the operation should use a strategy of rotating lunch periods to ensure that the constraint is never left idle.

3. Subordinate everything to the constraint. The optimal exploitation of the restriction is the primary concern. All other matters are of lesser importance.

4. Elevate the constraint. This refers to identifying methods to expand the total number of hours that are limited, such as obtaining additional hours.

5. Find a new constraint and repeat the process until the previous one is no longer in place. As the efficient exploitation of the limitation increases, it may no longer be a limitation as another process takes its place. Under those circumstances, the focus is redirected towards the newly identified limitation in the process.

2.30.3 Scheduling with the Theory of Constraints

As the effective utilization of the constraint improves, it may cease to be a limitation as another procedure supersedes it. In such situations, attention is shifted towards the recently found constraint in the process.

Drum. The term "drum of the system" pertains to the rhythmic and consistent pace at which manufacturing occurs. The master timetable for the operation is centered around the pace of throughput, as determined by the constraint. It is important to understand that after the speed at which the limitation operates has been determined, it is not beneficial to plan for greater production than the limitation can handle. Engaging in such action will only result in the accumulation of unfinished goods and might even diminish its overall efficiency.

Buffer. To ensure that the constraint always has the necessary inventory, a time buffer is commonly implemented before the constraint. The term "time buffer" is used to describe the duration for which the inventory in the buffer safeguards the constraint from any disturbances. Typically, multiple buffers exist in various systems, specifically a constraint buffer, an assembly buffer, and a shipping buffer. The constraint buffer is frequently used to account for the processing time in order to safeguard the buffer against unforeseen process variation. For instance, if a two-day time buffer is required, it means that upstream activities must finish processing and have the necessary material in the buffer two days prior to when it is actually needed. The first time buffer utilized is typically the aggregate duration from the initiation of raw material release to the point of reaching the constraint procedure. The assembly buffer often denotes the period between the release of raw materials and the stage where components that bypass the constraint process must be combined with components that do undergo the constraint process. The shipping buffer is the duration it takes for the material to be processed from the moment it leaves the constraint process until the final product is completed.

Rope. According to the analogy, production is directed toward the constraint for necessary processing by the rope's force. One way to achieve this is by deliberately releasing materials into the system at the right time, though this may also imply the use of a reactive replenishment system like a reorder point. In order to keep the buffer full, keep the constraint from running out of material, and prevent too much inventory from building up, the rope system schedules the release of raw materials into production. The limiting process's processing capacity largely determines it.

The analogy implies that the rope applies a push to guide production towards the constraint for necessary processing. Although it may imply the implementation of a reactive replenishment system, such as a reorder point, it is also feasible to do this by carefully introducing materials into the system at the optimal time.

The rope system synchronizes the schedule of raw material release into production to ensure the buffer is maintained, prevent the constraint from experiencing material shortages, and minimize excessive inventory buildup. The primary factor influencing it is predominantly dictated by the processing capacity of the constraining procedure.

- I-plant is a manufacturing process in which a single raw material is transformed into a singular end product. Processing often follows a linear path.
- A manufacturing facility where multiple subassemblies are combined to create a final assembly.
- The V-plant is capable of transforming a limited number of raw materials into multiple finished products.
- T-plant is a facility where many straight lines can divide into several assemblies.

The theory of limitations encompasses a systematic approach to facilitate the development and execution of organizational change. To begin, it is necessary to identify fundamental conflicts, which are subsequently confirmed by constructing a present reality tree. Once the negative consequences stemming from the main conflicts are recognized, a future reality tree is constructed to outline a strategic plan for resolving the issues. The ultimate crucial phase in the procedure entails constructing a tactical objective map that will delineate a strategy to achieve the envisioned future state.

Example 2.12: Parent X necessitates a single instance of both component Y and Z. Both Y and Z are executed on work center 20, which has a capacity of 40 hours available for processing. Component Y has a setup time of 1 hour and a run duration of 0.3 hours per piece. The setup time for component Z is 2 hours, while the run time is 0.20 hours per piece. Determine the quantity of Y_s and Z_s that can be manufactured.

Answer: Let x = number of Y_s and Z_s to produce

$\text{Tim e}_y + \text{Tim e}_z = 40$ hours

$1+0.3x+2+0.2x = 40$ hours

$0.5x = 37$ hours

$x = 74$

Therefore, work center 20 can produce 74 Y_s and 74 Z_s.

Example 2.13: In this Example, parent A consists of one B and two Cs. Both As and Bs are manufactured on work station 1, which has a weekly capacity of 40 hours. The Cs are manufactured on work station 2, which similarly has a weekly capacity of 40 hours.

Product	Setup Time (hours)	Run Time (hours/unit)
A	2	0.1
B	2	0.2
C	1	0.3

Determine the upper limit for the number of As, Bs, and Cs that should be manufactured per week, using the provided data.

Answer: The number of Bs produced should equal the number of As produced to avoid over production. Therefore, the number of Bs can be expressed in a formula as the number of As.

Work station 1

$\text{Time}_A + \text{Time}_B = 40$ hours

$2 + 0.1A + 2 + 0.2A = 40$

$0.3A = 36$

$A = 120$

Work station 1 has the capacity to produce enough As and Bs to make 120 As per week.

Work station 2

The number of Cs produced should be twice the number of A_s produced to avoid over production. Therefore, the number of Cs can be represented by 2 x A.

$Time_C = 40$ Hours

$1 + 2 \times 0.3A = 40$

$0.6A = 39$

$A = 65$

The capacity of work station 2 is limited to producing just 65 As every week, which is the constraint in this scenario. In order to prevent excessive production, it is recommended that work center 1 produces a total of 65 units of type A and 65 units of type B per week. Work station 2 should generate a total of 130 Cs every week, which is sufficient to yield 65 As.

In this particular instance, work station 1 will exhibit a significantly low level of usage. However, exceeding a production rate of 65 Bs per week will result in an accumulation of inventory. Additionally, work station 1 will have a shortage of resources due to work station 2, which can only create a maximum of 130 Cs per week.

2.30.4 Implementation

Orders that possess the requisite tooling, material, and capacity have a high likelihood of being finished punctually and can be authorized for production on the shop floor. Orders lacking essential items should not be released as they result in surplus work-in-process inventory and may disrupt the completion of other orders. Figure 2.19 illustrates the procedure for releasing an order.

Implementation is accomplished by giving manufacturing a shop order or schedule, which acts as their permission to start making the product. It is possible to assemble a shop package that contains the shop order and any required manufacturing information. The list may encompass any of the following options.:

- The shop order displays the shop order number, part number, name, description, and quantity. Technical illustrations.
- Bill of materials.
- Route papers display the prescribed processes, necessary equipment and accessories, required materials, and the estimated setup and run times.
- Material issue tickets are used to authorize manufacturers to obtain the necessary materials from retailers. These are also utilized for the purpose of billing the material in accordance with the shop order.
- Tool requisitions authorize the manufacturing department to withdraw the appropriate tooling from the tool crib.
- Individual job tickets for each operation to be executed. In addition to granting authorization for individual actions, they can also serve as components of a reporting system. The worker

has the ability to sign in and out of the job by utilizing the job ticket, which subsequently serves as a documented record of that particular task.

- Transfer tickets that authorize and instruct the transfer of work between different activities.

Currently, numerous manufacturing organizations employ a digital system that grants authorization for production based on a production schedule, rather than relying on the delivery of a shop order. The physical shop package is substituted with electronic access to the identical information. Online reporting is utilized as a substitute for physical tickets in order to document the movement of materials and report labor activities.

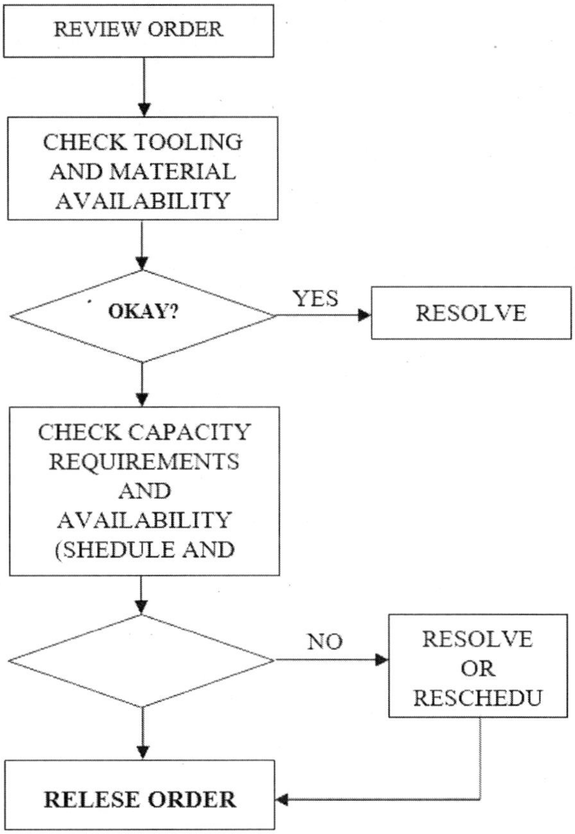

Figure 2.20 Order Release Process

2.30.5 Control

After work orders have been assigned to manufacturing, it is necessary to monitor and manage their progress. In order to manage progress, it is necessary to assess and compare performance against predetermined objectives. If there is a considerable deviation between the actual measurements and the expected outcomes, it is necessary to either modify the plans or implement corrective measures in order to align the performance with the original plan.

The primary goals of production activity control are to ensure timely delivery and optimize the utilization of corporate resources. In order to achieve specified delivery dates, a corporation must effectively manage the progress of orders on the shop floor, thereby exerting control over the lead time for orders. As previously mentioned in this chapter, the primary factor contributing to lead time is the queue. By effectively managing the queue, it is typically possible to meet delivery deadlines. Intermittent activities involve a wide range of items and order volumes, as well as various routings that require varying capabilities. It is exceedingly challenging to distribute the workload evenly across all the workstations in this particular setting. The existence of a queue is necessitated by the unpredictable nature of input and output.

In order to manage the queue and fulfill delivery obligations, it is necessary for production activity control to regulate the inflow and outflow of work at a work center. This is commonly referred to as input/output control. Establish the accurate precedence of instructions to be executed at each work center, also known as dispatching.

2.30.6 Input /Output Control

Production activity control is responsible for maintaining equilibrium in the movement of tasks between various work centers. The purpose of this is to maintain control over queue, work-in-process, and lead times. The input/output control system is a mechanism for overseeing and regulating the flow of tasks and the time it takes for them to be completed by overseeing and regulating the inflow and outflow of a facility. The system is engineered to equilibrate the input rate in hours with the output rate, ensuring precise control over both.

The rate at which input is received is regulated by the issuance of orders to the shop floor. Increasing the rate of input results in a rise in the queue, work-in-process, and lead times. The output rate can be regulated by adjusting the capacity of a work center, either by raising or decreasing it. Capacity fluctuations pose a challenge in manufacturing, but can be addressed through methods such as overtime or under time, workforce reallocation, and other strategies. Figure 2.24 visually represents the concept.

Input/output report In order to manage input and output, it is necessary to develop a strategy and establish a mechanism for comparing the actual outcomes with the anticipated ones. This information is displayed on an input/output report. Table 2.6 illustrates a sample of this report. The figures are expressed in normal hours.

Cumulative variance refers to the discrepancy between the projected total and the actual total during a specific period. The calculation is performed in the following manner:

Cumulative variance = previous cumulative variance + actual planned

Cumulative input variance week 2 = - 4+ 32

32 = - 4

The work to be done in hours is expressed by the backlog, which is the same as the queue. The following formula is used to calculate it:

Planned backlog for period 1 = previous backlog + planned input - planned output

$$= 32 + 38 - 40$$

$$= 30 \text{ hours}$$

The report indicates that the objective was to sustain a consistent level of production during each period and decrease the waiting time and processing time by 10 hours. However, both the input and output were lower than anticipated.

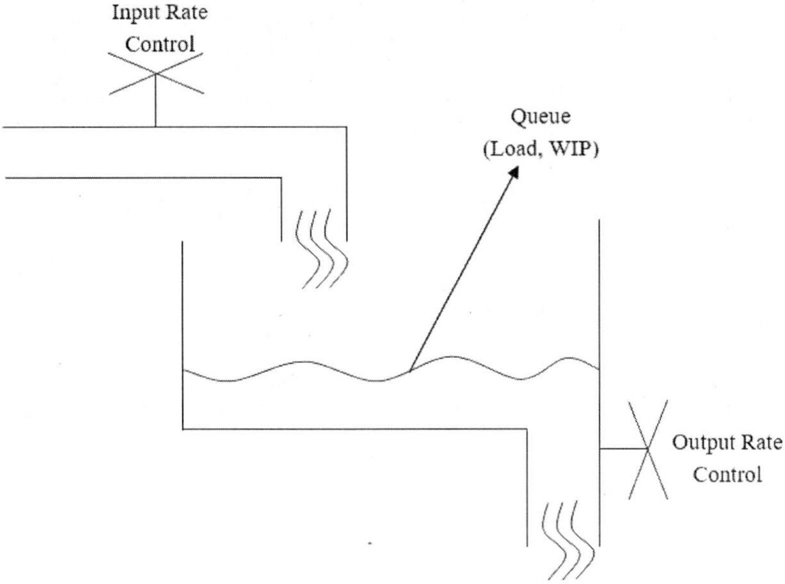

Figure 2.19 Input/ Output Controls

Work Center: 201

Capacity per period: 40 standard hours

Planned Input	40	40	40	40	40	200
Actual Input	32	36	44	44	36	192
Cumulative Variance	-8	-12	-8	-4	-8	-8
Period	1	2	3	4	5	Total
Planned Input	38	32	36	40	44	190
Actual Input	34	32	32	42	40	180
Cumulative Variance	-4	-4	-8	-6	-10	-10
Planned Backlog	32	30	22	18	18	22
Actual Backlog	32	34	30	18	16	20

Table 2.6 Input/ Output Report

The planned and actual inputs track the movement of tasks being assigned to the work center. The monitoring of the work center's performance is done through the comparison of planned and actual outputs. The planned and actual backlogs serve to monitor the queue and lead time performance.

Example 2.14

Week		1	2
Planned Input		45	40
Actual Input		42	46
Cumulative Variance			
Planned Output		40	40
Actual Output		42	44
Cumulative Variance			
Planned Backlog	30		
Actual Backlog	30		

Provide a comprehensive input/output report for weeks 1 and 2.

Answer

- Cumulative input variance week 1 = 42 45 = -3
- Cumulative input variance week 2 = -3 + 46 - 40 = 3
- Cumulative output variance week 1 = 42 - 40 = 2
- Cumulative output variance week 2 = 2 + 44 – 40 = 6
- Planned backlog week 1 = 30+ 45 - 40 = 35
- Planned backlog week 2 = 35 + 40 - 40 = 35
- Actual backlog week 1 = 30+ 42-42 = 30
- Actual backlog week 2 = 30+ 46 = 44 = 32

2.31 Production Reporting

Production reporting offers real-time input on the current activities taking place on the factory floor. PAC is able to keep accurate records of inventory levels, job progress, deficiencies, waste, material shortages, and other related information. The provision of this information is essential for production activity control to effectively determine appropriate priorities and address inquiries related to deliveries, shortages, and the current status of orders. This information is essential for manufacturing managers to make informed decisions on plant operations. The provision of this information is necessary for the payroll department to compute employees' remuneration.

It is necessary to gather, organize, and present information. The specific data collected is determined by the requirements of different departments. There is a range of methods used to acquire data. Occasionally, the operator utilizes an online system to promptly notify the initiation and conclusion of various activities, such as operations, orders, and movements. Alternatively, the operator, supervisor, or timekeeper may record this information on an operation reporting form that is part of the shop package. Reporting of inventory withdrawals and receipts is mandatory. After the data is gathered and documented, suitable reports are generated. The reports require different types of information, such as:

- Status of the order.
- Weekly data on the input and output of each department or work center.
- Exception reports provide detailed information on issues like as scrap, rework, and late shop orders.
- Current inventory status.
- Performance summaries are provided for order status, work center efficiency, department efficiency, and other related metrics.

2.32 Product Tracking

Production control often oversees the tracking of products and the traceability of lots. This refers to the procedure of tracing the origins of components and materials. Its practical application lies in the precise matching of colors in fabrics and paint, guaranteeing consumers that all units of the product maintain consistent color during manufacture and during the product's lifespan. Traceability may also be mandated in industries such as food, pharmaceutical, or aerospace to guarantee the safety of the product. If a product is found to be dangerous, the producer has the ability to track the origin of all materials and recall any final goods that utilized that specific batch. This procedure is highly detailed, involving the collection of information alongside other supply chain data, typically carried out by persons responsible for manufacturing activity control.

2.33 Measurement Systems

To effectively manage progress and make necessary adjustments to plans, it is essential to assess and compare performance against the anticipated objectives, as described before. Various performance measurement systems are available in the field of industrial activity control. The main objective of these metrics is to offer an impartial method of assessing performance and implementing necessary corrective measures. It is crucial to ensure that any measurement employed is consistent with the organization's overall performance measurement.

In addition to the previously listed evaluations of usage, effectiveness, output, shown capability, and input/output management, there are several more often employed metrics:

- **Comparison between the actual and projected lead time:** An analysis of the real time it takes to complete a process compared to the officially declared time frame.
- **Percentage of orders completed** on the scheduled date, without any delay or early completion.
- **Performance to schedule** refers to the comparison between the quantity and date of production and the master schedule.

UNIT SUMMARY

This unit provides a comprehensive examination of production forecasting, a critical component of effective resource management, production schedule planning, and client satisfaction. Companies can ensure the timely transportation of products and reduce costs by strategically synchronizing their supply chain operations through the process of predicting manufacturing requirements. Firms are able to make well-informed decisions regarding inventory levels, production capacities, and resource allocation by comprehending demand patterns and employing both subjective and objective prediction techniques. Precise prediction not only increases operational effectiveness but also boosts customer contentment by minimizing the time between order and delivery and avoiding instances where inventory is depleted. Scheduling, an essential component of production management, seeks to maximize the usage of resources, reduce periods of inactivity, and improve output. By implementing the processes of loading, sequencing, and monitoring, organizations may enhance the coordination of production activities, resulting in seamless operations and punctual delivery of goods. Advanced Planning and Scheduling (APS) systems, in conjunction with techniques such as the Theory of Constraints, provide effective tools for improving production schedules and resource allocation. In addition, efficient staff scheduling maximizes the use of worker resources, leading to improved operational efficiency and organizational success. Production forecasting and scheduling are essential elements of industrial operations, playing a crucial role in enhancing efficiency, minimizing expenses, and sustaining a competitive edge in the ever-changing marketplace. All these topics are discussed in this unit.

EXERCISES NO: -2

Multiple Choice Questions

1. Which of the following best defines production forecasting?
   a. Estimating the cost of production
   b. Predicting future demand for products
   c. Analyzing historical production data
   d. Determining optimal inventory levels
2. What is the strategic role of forecasting in the supply chain?
   a. Reducing production costs
   b. Minimizing inventory levels
   c. Improving customer satisfaction
   d. Optimizing transportation routes
3. What is the primary time frame considered in production forecasting?
   a. Short-term
   b. Medium-term
   c. Long-term
   d. All of the above

4. Which of the following describes a qualitative method of forecasting?
    a. Moving Average
    b. Exponential Smoothing
    c. Delphi Method
    d. Time Series Analysis
5. Quantitative forecasting techniques primarily rely on:
    a. Expert judgment
    b. Historical data
    c. Market research
    d. Consumer surveys
6. Which of the following is an example of a quantitative forecasting technique?
    a. Market testing
    b. Jury of Executive Opinion
    c. Regression Analysis
    d. Sales force composite
7. Qualitative forecasting techniques are often used when:
    a. Historical data is unavailable
    b. Future trends are easily predictable
    c. The market is stable
    d. Accuracy is crucial
8. Which technique involves analyzing the relationship between a dependent variable and one or more independent variables?
    a. Exponential Smoothing
    b. Time Series Analysis
    c. Regression Analysis
    d. Moving Average
9. Qualitative forecasting techniques are best suited for:
    a. Short-term forecasts
    b. Stable demand patterns
    c. Uncertain or rapidly changing environments
    d. Highly accurate predictions
10. Which forecasting method uses judgment and intuition rather than mathematical calculations?
    a. Exponential Smoothing
    b. Trend Projection
    c. Jury of Executive Opinion
    d. Moving Average
11. Which technique involves projecting past data into the future based on historical patterns?
    a. Exponential Smoothing

b. Time Series Analysis
c. Delphi Method
d. Market Testing

12. When might quantitative forecasting techniques be preferred over qualitative techniques?
    a. When historical data is unreliable
    b. When expert opinions are readily available
    c. When the market is highly volatile
    d. When accuracy is not critical
13. How is forecast accuracy typically measured?
    a. Mean absolute deviation
    b. Standard deviation
    c. Variance
    d. Mode
14. What is the primary objective of scheduling?
    a. Maximizing employee satisfaction
    b. Minimizing production costs
    c. Optimizing resource utilization
    d. Balancing inventory levels
15. What does APS stand for in the context of scheduling systems?
    a. Advanced Production Software
    b. Agile Planning System
    c. Advanced Planning and Scheduling
    d. Automated Production System

Answers of Multiple-Choice Questions

1 b, 2 c, 3 d, 4 c, 5 b, 6 c, 7 a, 8 c, 9 c, 10 c, 11 b, 12 c, 13 a, 14 c, 15 c

Short and Long Answer Type Questions

1. In the production planning system, what four tasks are carried out by the master production schedule (MPS)?
2. What tasks does the MPS carry out in the interim between production and sales?
3. Discuss the strategic role of productiön forecasting in supply chain management.
4. Compare and contrast qualitative and quantitative forecasting methods in terms of their advantages and limitations.
5. Explain the steps involved in developing a production forecast using quantitative methods.
6. How do businesses utilize historical data and market trends in production forecasting?
7. Discuss the challenges associated with production forecasting in volatile markets and how businesses can mitigate them.

8. Explain how the Theory of Constraints helps identify and address bottlenecks in production scheduling.
9. Discuss the challenges associated with employee scheduling and strategies for optimizing workforce utilization.
10. How do emerging technologies like artificial intelligence and predictive analytics impact production scheduling processes?

KNOW MORE

REFERENCES

[1] *GROOVER, M. P., Fundamentals of Modern Manufacturing: Materials, Processes, and Systems, 5th ed., John Wiley & Sons, Inc., Hoboken, NJ, 2013.*

[2] *GROOVER, M. P., Work Systems and the Methods, Measurement, and Management of Work, Pearson/Prentice Hall, Upper Saddle River, NJ, 2007.*

[3] *HARRINGTON, J., Computer Integrated Manufacturing, Industrial Press, Inc., New York, NY, 1973.*

[4] *Hill, T. (1993) Manufacturing Strategy: The Strategic Management of the Manufacturing Function, 2nd ed. Open University, Macmillan, London.*

[5] *Patel, P. M. (1993) Quality and productivity improvement through effective management of control chart and analysis of statistical data. Quality Forum 19, 71-79.*

[6] *Transactions on Automation Science and Engineering, vol. 14, no. 3, 2019, pp. 567-580. Valeo Automotive Components (1995) The Valeo Production System, Swansea, Wales (UK).*

-----------------*---------*

3 BREAK-EVEN ANALYSIS & AGGREGATE OPERATIONS PLANNING

UNIT SPECIFICS

The following topics are covered in this unit:

- Concept on Price
- Value perception of customer
- Company and Product costs
- Types of costs
- Break-Even Analysis: Introduction, Break-even analysis charts, Breakeven analysis for process, plant and equipment selection.
- Overall Marketing Strategy, objectives and Mix
- Aggregate production planning,
- Adjusting capacity to meet the demand,
- Demand management,
- Hierarchical and collaborative planning,
- Aggregate planning for services.

All the topics are discussed in a lucid manner along with their engineering applications. For student's practice, a moderate number of multiple-choice questions, short and long answer type questions and numerical problems are also given.

RATIONALE

This unit, which is designed for diploma engineering students, explores a variety of critical components of business and marketing. It commences by delving into the complex relationship between price and value from the customer's perspective, underscoring the significance of comprehending consumer perceptions in the development of pricing strategies. Moving forward, students are introduced to the complex network of costs that are inherent in both company operations and product development. This network includes a variety of categories, including overhead, variable, and fixed costs. The unit also provides students with the necessary tools to evaluate profitability and make informed decisions by further elucidating the fundamental concepts of break-even analysis. Furthermore, it offers a comprehensive understanding of the development of a strong marketing strategy, particularly in the context of engineering initiatives, by clarifying objectives and strategies. Students acquire the ability to seamlessly incorporate these business principles into their engineering pursuits through practical applications and lucid explanations.

The manufacturing planning and control (MPC) method is explained in this chapter. It starts by talking about the whole system and then moves on to some specifics of planning production. After that, talk about master ordering, planning for material needs, capacity management, controlling production activities, and buying.

Making plans for production is the first part of a method for planning and controlling production. The time frame for planning is generally one year. The minimum horizon is based on how long it takes to get the supplies and make the product. There isn't a lot of information. Most of the time, plans are made for groups of goods that are made in similar ways or that share a unit. The production plan is a part of the sales and operations planning process. This is an executive-level planning process that involves making trade-offs between different company departments or roles.

You can make a production plan in one of three main ways: chase, leveling production, or hybrid. Each has pros and cons when it comes to operations and costs. Manufacturing management has to choose the best mix of these basic plans so that total costs are kept as low as possible and customer service levels are kept high.

Furthermore, the unit offers a wide variety of queries and problems to facilitate the application of students' newly acquired knowledge. It also provides plenty of opportunities for assessment.

PRE-REQUISITES

- Fundamental Knowledge of any Manufacturing Industry Layout
- A course on Industrial Engineering
- Fundamental Knowledge of Marketing

UNIT OUTCOMES

After understanding this chapter readers will be able to

U3-O1: Discuss pricing's importance in today's fast-changing world.

U3-O2: Examine the significance of product and company costs in the determination of prices.

U3-O3: Understand the concept of Breakeven analysis and chart.

U3-O4: Discuss the Production planning system in Industry

U3-O5: Understand the concept of Capacity and demand Management

Unit-3 Outcomes	EXPECTED MAPPING WITH COURSE OUTCOMES (1- Weak Correlation; 2- Medium correlation; 3- Strong Correlation)				
	CO-1	CO-2	CO-3	CO-4	CO-5
U3-O1	3	1	0	2	3
U3-O2	3	2	1	3	3
U3-O3	2	3	2	3	3
U3-O4	3	3	3	2	1
U3-O5	3	3	2	2	2

3.1 Concept on Price

It is commonly believed that cost is a fact and price is a policy. In its most basic form, price refers to the total of money that is charged for an artifact or service. In a broader meaning, price is the aggregate of all the values that consumers forfeit in exchange for the advantages of possessing or employing a product or service. Throughout history, the primary determinant of consumer preference has been price. Non price factors have become increasingly significant in recent decades. Nevertheless, price continues to be a critical factor in determining a company's profitability and market share.

Revenue is generated solely by price; all other components represent expenses. Additionally, price is one of the most adaptable components of the marketing formula. Prices are subject to rapid modification, in contrast to product quality and channel commitments. Simultaneously, pricing is the most significant challenge encountered by numerous marketing administrators, and numerous organizations fail to effectively manage pricing. One common issue is that companies are too eager to reduce prices in order to secure a transaction, rather than persuading consumers that the superior value of their product warrants a higher price. Other prevalent errors include pricing that is excessively cost-oriented rather than customer-value-oriented, and pricing that fails to consider the remaining components of the marketing mix.

Pricing is perceived as a significant challenge by certain managers, who would concentrate on the other components of the marketing mix. Nevertheless, astute managers regard pricing as a critical tactical instrument for generating and retaining consumer value. Pricing directly influences an organization's profitability. Increasing prices by a minor percentage can result in a substantial rise in productivity. Price is a critical component of a company's total value proposition, as it is instrumental in the development of client value and the establishment of client relationships. "Instead of running away from pricing," according to the specialist, "savvy marketers are embracing it."

3.1.1 Key considerations for price setting

Author Comment: *Selecting the appropriate item is one of the marketer's most challenging responsibilities. Multiple factors are involved. However, the process of discovering and executing the appropriate price plan is crucial for achieving success.*

The corporation will choose a value that is neither too high to generate demand nor too low to generate profit. Figure 3.1 provides a concise overview of the key factors to consider when determining the price. The maximum price of the product is determined by how customers perceive its value. If clients think that the price exceeds the product's value, they will refrain from buying it. Product costs establish the minimum prices. Should the corporation set the product price below its costs, it will negatively impact corporate profitability. When determining its price, the company needs to take into account several internal and external aspects, such as its overall marketing policy and mix, the characteristics of the market and demand, and the tactics and prices of competitors.

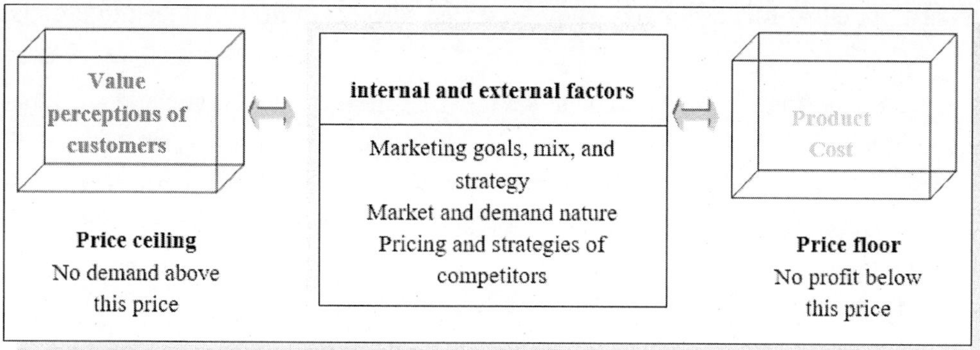

Figure 3.1 Summarizes the Key Factors that Go into Setting Prices

Note: - If buyers think that the charge of a product exceeds its value, they will not make a purchase. When a firm sets amount for a product that is lower than its cost, it will see a decrease in earnings. Therefore, the ideal pricing strategy is one that provides both value to the client and profit to the company.

3.2 Value perceptions of customers

Ultimately, the determination of whether a product's pricing is appropriate rests with the client. When making pricing selections, it is essential to begin by considering the value that customers perceive in the product or service. Customers engage in a transaction where they give up something valuable (the price) in order to obtain something valuable (the benefits of owning or utilizing the goods). Customer-centric pricing is achieved by comprehending the extent to which consumers value the profits they derive from a product and establishing a price that accurately reflects this value.

3.2.1 Pricing Based on Value

Effective pricing begins with a comprehensive comprehension of the value that a product or service generates for customers. worth-based pricing relies on the buyers' subjective assessment of worth rather than the seller's production cost to determine the price. Value-based pricing refers to the approach where the marketer is unable to determine the price of a product and develop a marketing program without considering its value. Pricing is taken into account alongside the other elements of the marketing mix prior to establishing the marketing campaign.

Pricing Based on Value: Pricing based on purchasers' subjective assessment of worth rather than the seller's production expenses.

Figure 3.2 presents a comparison between value-based pricing and cost-based pricing. Cost-based pricing is determined by the expenses incurred in producing a product. The corporation engages in the process of conceptualizing and creating a product that it deems to be of high quality. It then calculates the expenses associated with manufacturing the product and establishes a price that covers these costs also includes a desired level of profit. Marketing must thereafter persuade purchasers that the product's worth at that particular price is sufficient to justify its acquisition. If the price is deemed excessive, the corporation must opt either reduced markups or decreased sales, both of which lead to unsatisfactory earnings.

Value-based pricing operates in the opposite direction. Initially, the organization evaluates the requirements and the way customers perceive the value. Subsequently, it establishes its desired pricing by taking into account the customers' subjective assessment of the product's worth. Decisions about allowable costs and product design are determined by the targeted value and pricing. Consequently, the process of determining pricing starts by examining the requirements and subjective assessments of consumers, and the price is established to align with consumers' perceived worth.

Keep in mind that "good value" is not the same as "low price." For instance, some people think that the pricey Rs. 175,000 Bentley Continental GT is a good deal. Because a car those costs Rs. 175,000 is not actually expensive, but rather a great deal. It takes 160 hours to build each Bentley GT by hand, which is an old-fashioned way of making cars. It takes 18 hours for craftsmen to stitch the perfectly joined leather of the GT's steering wheel together. That's almost as long as it takes to put together a whole VW Golf.

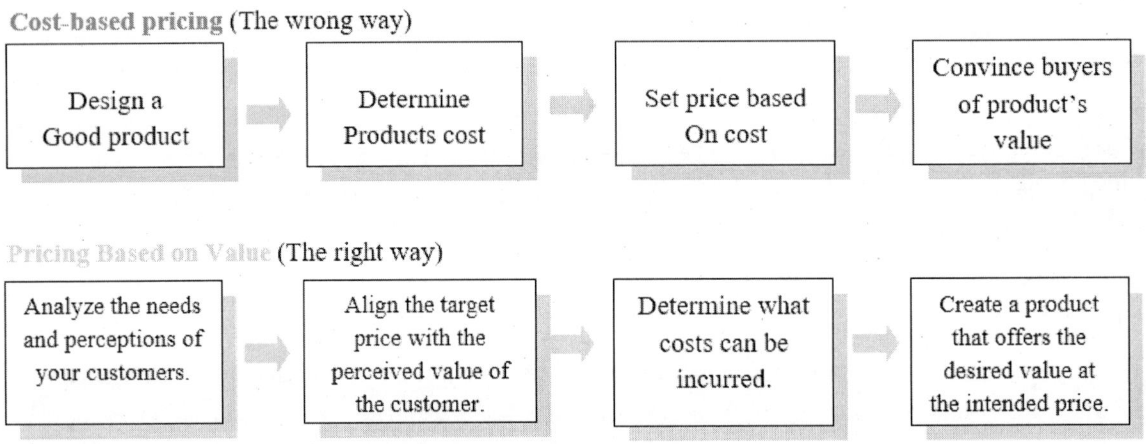

Figure 3.2 Value-based Pricing v/s Cost-based Pricing

These are the impressive results: Dashboard and doors are covered in a mirror-like wood veneer. The floor pedals are carved from aluminum, the window and seat toggles are made of metal instead of plastic, and every air vent is chromed to perfection. All of this adds up to a cabin that looks like it belongs in a Rs. 300,000 car and an engine that sounds like it belongs in a Rs. 200,000 car. The car has brilliantly combined... technological sophistication The GT is a good deal, as I said. You can ask anyone on the extended list of people who are waiting. Right now, you have to wait six to seven months before you can take home your very own GT.

If a business uses value-based price, it needs to know how much different competitive offers are worth to buyers. But it's not always easy for businesses to figure out how much charge buyers will put on their products. For instance, it's not too hard to figure out how much the food at a fancy restaurant costs. But it's hard to put a number on things like environment, conversation, taste, relaxation, and status etc. And these ideals will be different for every customer and every situation.

Customer will still use these values to judge the price of a product, so the company needs to work to measure them. People are sometimes asked about how much would they pay for a basic product plus all the extras that come with it. Or, a business might test how valuable different product offers are seen to be by doing experiments. An old Russian saying says “there are two fools in every market,one who asks too much and one who asks too little”. The company will lose sales if the vender charges more than what the buyers think the item is worth. The seller's goods sell very well if they cost less. But they bring in less money than they would if they were priced at what people think they are worth.

Let us now analyze two forms of value-based pricing: good-value pricing and value-added pricing.

3.2.2 Good-Value Pricing

Over the past years, marketers have observed a significant change in consumer perceptions regarding pricing and quality. Several organizations have adjusted their pricing strategies to align with evolving economic circumstances and consumer price perceptions. Marketers have increasingly embraced good-value pricing methods, which involve providing a balanced combination of quality and good deal at a reasonable price.

Most of the time, this has meant making cheaper versions of well-known brand-name goods. Because of the tougher economy and people's tendency to spend less, fast food places like McDonald's give "value menus." Armani has a design line called Armani Exchange that is less expensive and more casual. In other news, Tata Motors just released the Tata Nano, a cheap car that starts at about Rs. 100,000.

Sometimes, good-value pricing has meant redesigning well-known names to provide better quality for the same price or better quality for less money. Tata Motors, for example, recently released the Tata Tiago, a "compact sedan" that costs less than Rs. 700,000. Some businesses are so successful that they can give less value at very low prices. For example, people who fly with the low-cost European plane Ryanair won't get many freebies, but they will love how cheap the tickets are.

Everyday low pricing (EDLP) is a type of good value price that is important at the mall level. For EDLP to work, prices must stay low all the time, with few or no short-term price cuts. High-low pricing, on the other hand, means charging higher prices every day but having frequent sales to temporarily lower prices on certain items. EDLP has replaced high-low pricing in stores that sell everything from gadgets to furniture in the last few years. Wal-Mart is the king of EDLP because it pretty much created the idea. Everyday small prices on everything Wal-Mart sells, besides a few things that go on sale every month. In the same way, Big Bazaar has the best deals by saying, "Isse sasta aur kahaan?" (Where else can I get it for less?)

3.2.2.1 A Case study of Ryanair Air Lines

In these tough times for air travel, it's hard for the big companies to decide how to set their prices. Pricing tactics are very different. Southern, JetBlue, Frontier, and Skybus are some of the airlines that give simple flights at very low prices. Others, like Virgin, Silver jet, Eos, and Singapore Airlines, give more comfort and charge more to match. However, most airlines still haven't figured out how to price their tickets, which makes air travelers usually unhappy when talking about ticket prices.

One air line seems to have arisen up with a crazy new way to charge for flights that customers will love: make them free! An expert for Business 2.0 magazine says:

Yes, Michael O'Leary wants to make flying free. He is the CEO of Ryanair in Ireland, which is the most profitable airline in Europe. Free doesn't mean "not regulated," but it does mean "no cost." I promise that by the end of the decade, "more than half of our travelers will fly free." It's interesting that not many experts think his guess is too far-fetched: Ryanair already gives free tickets to a quarter of its customers.

Ryanair has become one of the most popular airlines in Europe, even though it doesn't offer free trips. It took 42.5 million people to more than 100 places in Europe last year. The company made 33% more money and sold more than Rs. 3.6 billion, which is 32% more than the previous year. Southwest is the U.S. low-cost leader with an average price of Rs. 107, but Ryanair's net margins are 21%, which is three times Southwest's 7%. There will likely be rough times ahead for the airline business, with fuel prices going up, the economy going down, and other problems. However, Ryanair looks like it will be able to handle the storm.

How do you do it? Ryanair's low prices make even Southwest, which is known for being cheap, look like they spend a lot of money without thinking. The Irish airline also charges extra for almost everything except tickets. This includes everything from checking bags to putting ads on the back of your seat. "[Ryanair] thinks like a retailer and charges for absolutely every little thing, except the seat itself," says an analyst. "Imagine the seat as akin to a cell phone: It comes free, or nearly free, but its owner winds up spending on all sorts of services."

Southwest's low-cost approach is what Ryanair is trying to copy. CEO O'Leary went to Dallas in 1991 to meet with Southwest officials and see what he could learn. At the time, Ryanair was just another bad European airline. The Irish company had to completely change how it did business as a result. To save money, Ryanair started using only one type of plane, the good old Boeing 737, like Southwest did. Like Southwest, it started to focus on smaller, secondary airports and let passengers choose where to sit.

But Ryanair has since made Southwest's plan for low prices even better. O'Leary is a total fanatic when it comes to cutting costs. "We want to be known as the Wal-Mart of flying," he adds. Ryanair is always looking for new ways to save money, just like the big store. For example, they got rid of seat-back bags to save money on cleaning and weight. People who take the train gain from these savings because fares are lowered. Ryanair also sells more than 98% of its tickets online, which cuts down on the costs of running the business and paying travel agents. The flight teams even buy their own uniforms, and the people who work at headquarters bring their own pens.

The company that is always keen to save money also charges customers for almost every extra service they use. On Ryanair, you can't get a free drink or bag of pretzels. The airline was the first to charge customers for foodstuff while they were in the air, which brings in tens of millions of dollars every year. It was also the first to charge a fee to check bags, which it made up for by lowering ticket prices by the same amount. By cutting down on fuel and handling costs, that move saved tens of millions of dollars. Just as aggressively, the company is looking for new ways to make money.

Some of the companies that Ryanair has put ads for on its planes are Vodafone Group, Jaguar, and Hertz. Soon, ads will also be shown to every person when their seat back trays are raised. Flight workers try to sell everything to the people who are stuck in the airborne, from scratch-off cards to digital cameras and perfumes. If you get off at a strange airport, Ryanair sells bus and train tickets into city. Ryanair makes more money from its website, which gets 15 million unique users a month. When this business sells Hertz rental cars, ski packages, hotel rooms and trip insurance, it gets paid a fee. These other kinds of income rose 36% to Rs. 332 million last year. An expert in the field says, "Every chance they get, Ryanair tries to squeeze just that little bit of extra margin out of its passengers."

Customers aren't complaining about Ryanair's ridiculously cheap prices, especially since they can choose not to buy anything else. A lot of them are also fun to use or make life easier:

Ryanair plans to add gambling to its flights soon so that customers can pass the time even more. The airline will get a small cut of each bet. Ryanair could make twice as much money in ten years if O'Leary is right about gaming, but he's not done yet. He also thinks about a time when the company will be able to charge people to use their cell phones at 35,000 feet. He also wants to work with people who run airport parking lots and food stands to get a bigger share of the money that people spend while they're waiting for or arriving at his planes.

The Business 2.0 analyst comes to this conclusion: "When you add it all up—constantly cutting costs on the operations side along with new ways to get more money from each passenger—O'Leary's plan to give away half of Ryanair's seats by 2010 starts to look pretty reasonable." Ryanair may make an impression as if you're going in a subway car instead of an airplane, but the prices are too good to pass up. U.S. airlines that are having trouble making ends meet should pay attention: letting people fly for free could be truly freeing. Ryanair doesn't think the sky is the end.

Sources: Quotes and excerpts from or adapted from Matthew Maier, "A Radical Fix for Airlines: Make Flying Free," Business 2.0, April 2006, pp. 32-34, and Kerry Capell, "Wal-Mart with Wings," BusinessWeek, November 27, 2006, pp. 44-46. Also see "Ryanair Offers Gambling Web Site," Associated Press, November 1, 2006; Will Sullivan, "Flying on the Cheap," U.S. News & World Report, March 26, 2007, p. 47; "Ryanair Offers Ad Space in Planes," Marketing, February 28, 2007; Mark Tatge. "Nightmare at 30,000 Feet," Forbes, June 4, 2007, p. 56; Kevin Done, "Ryanair Warns of Perfect Storm Damage," Financial Times, February 5, 2008, p. 17; Kerry Capell, "Fasten Your Seatbelt, Ryanair," BusinessWeek, February 18, 2008, p. 16; and www.southwest.com and www.ryanair.com, accessed November 2008.

3.2.3 Value-Added Pricing

Setting charges based on value is more than just getting what clients want to pay or being cheap to compete. When it comes to marketing, one of the hardest things is to give the company the pricing power it needs to avoid price war and defend higher prices and margins. For a company to have more pricing power, it needs to keep or improve the value of what it sells. This is especially true for companies that sell common goods, since these goods don't vary much and there is a lot of price competition.

A lot of businesses use value-added pricing methods to get more control over their prices. Instead of lowering their prices to compete, they add features and services that make their products be noticeable, which allows them to charge more. A Look at this example:

In Kerala, India, it rains almost every day during the rainy season for three months. For 147 years, most people in the area stayed safe with a Stag umbrella from the famous Ebrahim Currim& Sons. Like Ford's Model T, the basic Stag was strong, cheap, and could be any color as long as it was black. But by the end of the 20th century, cheaper goods from China were a threat to the Stag. In response, Stag lowered its costs and scrimped on quality. It was a bad idea because the brand started losing money for the first time since the 1940s. In the end, though, the company got smart. It gave up the price war and promised to make the product better. Surprisingly, sales of the better Stag umbrellas went up even though they cost more. So the business began to come up with new ideas. It came out with designer umbrellas in cool colors and funky designs because Indian guys are becoming more aware of the need for fresh air. Teenagers and young adults loved them. Then, Stag made umbrellas with a powerful flashlight built in for people who walk at night on dark roads and models with pre-recorded music for people who like music. There is a form of Stag's Bodyguard with glare lights, emergency blinkers, and an alarm for women who walk alone at night. People are willing to pay up to a hundred percent more for the new goods. The Stag brand is now making money again thanks to the new value-added approach. The big old black Stags still show up on the streets of Kerala during the rainy season in June, but they cost 15% more than the imported ones.

The Stag example demonstrates once more that customers are driven by the value, they receive in relation to the price they pay, rather than solely by the price itself. "If consumers believed that the optimal choice was solely determined by the amount of money saved, we would all be shopping in a single large discount store," states a pricing specialist. "Customers desire value and are prepared to compensate for it." Shrewd marketers set the prices of their items appropriately.

3.3 Company and Product Costs

Author comment: Cost establishes the floor for price; however, the objective is not always to reduce costs. Bentley automobiles are an example of a company that invests in higher costs in order to claim higher prices and margins. Management of the disparity between costs and prices is crucial, as it determines the company's profitability in relation to the value it provides to its customers.

Unlike customer-value perceptions, which establish the price ceiling, costs establish the floor for the price that the company can charge. Cost-based pricing is the process of determining prices by taking into account the costs of production, distribution, and sale of the product, as well as a reasonable return on investment for the effort and danger involved. The pricing policy of a company may be significantly influenced by its costs.

Indigo Airlines, Makro-Habib, and Dell are among the organizations that strive to establish themselves as "low-cost producers" within their respective sectors. Companies with reduced costs have the ability to establish lower prices, which leads to increased sales and profits on the basis of smaller margins.

Nevertheless, other organizations deliberately incur higher expenses in order to establish higher charges and margins. For instance, the cost of producing a Bentley that is "built by hand" is higher than that of a Maruti 800. However, the astronomical Rs. 175,000 price is justified by the superior quality that results from the increased costs. Management of the disparity between costs and prices is crucial, as it determines the company's profitability in relation to the value it provides to its customers.

3.3.1 Cost-based pricing

Establishing prices by taking into account the costs of production, distribution, and sale of the product, as well as a reasonable return on investment and risk.

- **Fixed costs (overhead):** Costs that are constant regardless of the quantity of production or sales.
- **Variable cost:** cost that is directly proportional to the production volume.
- **Total cost:** For any particular production level, the sum of the fixed and adaptable costs.

3.3.2 Types of Costs

There are two distinct types of costs that a company incurs: fixed and variable. Fixed costs, which are also referred to as overhead, are expenses that remain constant regardless of the level of production or sales. For example, regardless of the company's output, it is necessary to pay the monthly expenses for rent, heat, interest, and executive salaries. The level of production directly influences variable costs. Computer processors, wires, plastic, packaging, and other inputs are all included in the cost of each PC manufactured by Hewlett-Packard. Each product produced typically incurs identical expenses. They are referred to as variable because their total quantity fluctuates in accordance with the quantity of units manufactured. The sum of the fixed and adaptable costs for any given level of production is known as the total costs. Management aims to establish a price that will at least cover the total production costs at a specific production level.

The company must exercise caution with respect to its expenditures. The company will be at a competitive disadvantage if it is required to charge a higher price or generate a lower profit than its competitors in order to compensate for the increased costs associated with producing and selling its product.

3.3.3Costs at various Levels of Production

Management must be aware of the fluctuations in costs that occur at various production levels in order to make informed pricing decisions. For instance, let us assume that Casio has constructed a facility capable of producing 1,000 calculators per day. Figure 3.3A illustrates the typical short-run average cost curve (SRAC). It demonstrates that the cost of a calculator is excessive if Casio's factory produces only a few units per day. However, the average cost decreases as production increases to 1,000 calculators per day. this is due to the fact that fixed costs are distributed across a greater number of units, with each unit bearing a lesser portion of the fixed cost. Casio may endeavor to manufacture over 1,000 calculators per day; however, the plant's inefficiency will result in an increase in average costs. Workers are required to wait for machines, which are more susceptible to malfunctions, and they frequently interfere with one another.

Casio should contemplate the construction of a larger facility if it is confident in its ability to sell 2,000 calculators per day. The facility would implement more efficient machinery and work arrangements. Additionally, the long-run average cost (LRAC) curve (Figure 10.3B) illustrates that the unit cost of producing 2,000 calculators per day would be less than that of producing 1,000 units per day. Figure 10.38 indicates that a plant with a capacity of 3,000 would be even more efficient. However, a production plant that produces 4,000 units per day would be less efficient due to the growing number of diseconomies of scale, such as the burden of managing an excessive number of employees and the slowdown caused by documentation. The optimal capacity for a production plant to be constructed is a 3,000-daily plant, as illustrated in Figure 10.3B, provided that the demand is sufficient to sustain this level of production.

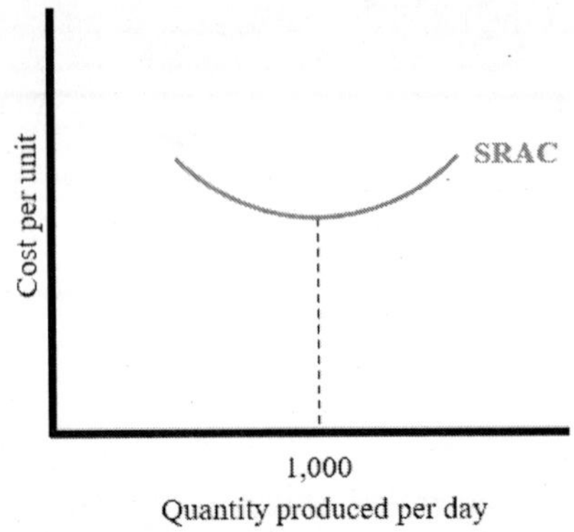

Figure 3.3 Cost Dynamics over a Different Size Plant

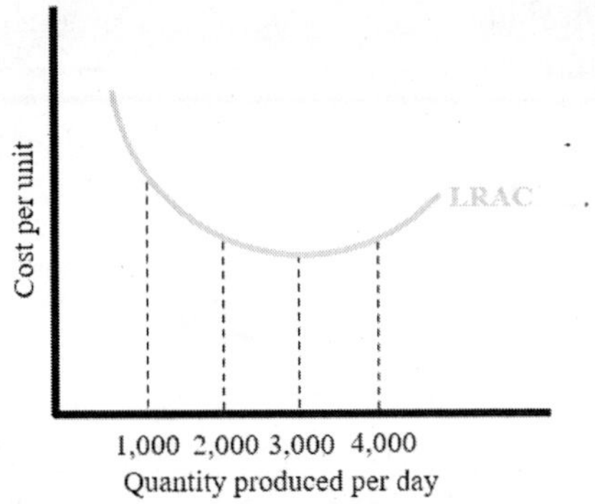

Figure 3.4 Cost Dynamics in a Fixed-Size Facility

3.3.4 Costs as a Function of Production Experience

Assume that Casio operates a manufacturing facility that generates 3,000 calculators daily. Casio acquires expertise in the production of calculators as it accumulates experience. Workers develop a greater understanding of their equipment and acquire alternatives. The work becomes more organized as a result of practice, and Casio develops more efficient production processes and equipment. Casio achieves economies of scale and increases its efficiency as its volume increases. Consequently, the typical cost decreases as production experience accumulates. **Figure 3.4** illustrates this. Consequently, the average cost of manufacturing the initial 100,000 calculators is Rs. 500 per calculator. The average cost has decreased to Rs. 450 after the company has manufactured the initial 200,000 calculators. The average cost is Rs. 350 after the accumulated production experiençe doubles again to 400,000. The experience curve (or the learning curve) is the term used to define the decrease in the average cost that occurs as production experience accumulates.

Experience curve (learning curve): **The decrease in the average per-unit production cost that results from the accumulation of production experience**

The company is significantly impacted by the existence of a downward-sloping experience curve. The company's unit production cost will decrease, and it will decrease at a quicker rate if the company produces and sells more during a specific period. However, the market must be prepared to purchase the increased output. In order to capitalize on the experience curve, Casio must establish a substantial market share at the outset of the product's life cycle. This implies the implementation of the subsequent pricing strategy: The calculators should be priced at a low level, which will result in an increase in sales and a decrease in costs as a effect of the company's increased experience. Subsequently, the company can further lessen its prices.

Companies have implemented successful strategies that revolve around the experience curve. For instance, Bausch & Lomb solidified its position in the soft contact lens market in the United States by expanding its single S of Lens facility and utilizing computerized lens design. Consequently, its market share increased consistently to 65 percent.

But focusing only on cutting costs and taking advantage of the experience curve won't always work. There are some major problems with experience-curve prices. Because the prices are so low, the goods might look cheap. It's also thought that rivals are weak and won't fight back by lowering their prices to match the company. Lastly, while the company is building up its business with one technology, a rival may discover a cheaper one that lets it start at a lower price than the market leader, who is still using the old experience curve.

3.4 Cost-Plus Pricing

Cost-plus pricing is the most basic technique of pricing, where a predetermined markup is added to the product's cost. Construction firms, for instance, provide job proposals by calculating the overall project expenditure and incorporating a fixed percentage increase for profit. Professionals such as lawyers, accountants, software consultants, and others usually determine their prices by applying a customary markup to their expenses. Certain vendors inform their clients that they will impose a fee in addition to the actual cost, known as a stated markup. For instance, aerospace businesses adopt this pricing strategy when dealing with the government.

Cost-plus pricing is the addition of a predetermined markup to the cost of a product.

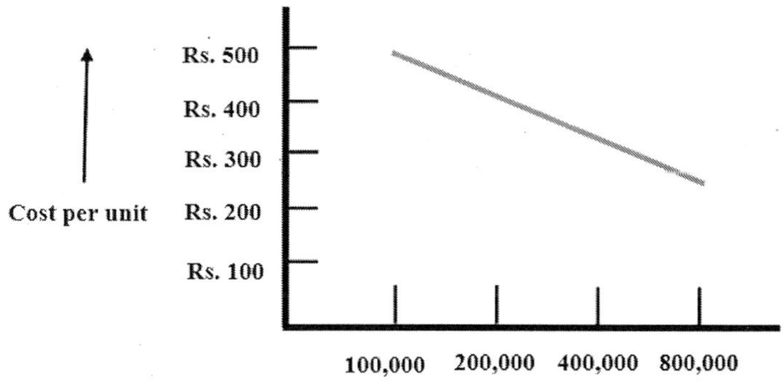

Figure 3.5 Cost Per Unit in Relation to Total Production. The Experience Curve

To exemplify the concept of markup pricing, consider a premium manufacturer of pure fruit juice that has the following costs and projected sales for its 1-liter packaging:

Variable cost	Rs. 10
Fixed costs	Rs. 300,000
Expected unit sales	50,000

The manufacturer's cost per liter pack can be calculated using the following formula:

$$\text{Unit Cost} = \text{Variable Cost} + \frac{\text{Fixed cost}}{\text{Unit sales}} = \text{Rs. } 10 + \frac{\text{Rs.}300{,}000}{50{,}000} = \text{Rs. } 16$$

Assuming the firm aims to achieve a 20 percent profit margin on sales. The manufacturer's markup price is determined by the following formula:

$$\text{Markup Price} = \frac{\text{Unit Cost}}{(1 - \text{Desired Return on Sales})} = \frac{\text{Rs. } 16}{1 - 2} = \text{Rs. } 20$$

The company that made them would get Rs. 20 per pack from sellers and make Rs. 4 per unit. The sellers will then charge more for the juice pack. They will raise the price of the pack to Rs. 40 (Rs. 20 plus 50% of Rs. 40) if they want to make 50% of the sale price. This amount is the same as a 100% markup on the cost (Rs. 20/Rs. 20).

Does it make sense to set prices with normal markups? In most cases, no. If you set your prices without taking demand or competitor prices into account, you probably won't get the best deal. Markup price is still common for many reasons. For starters, sellers are sure about prices but not so sure about demand. By linking the price to the cost, sellers make pricing easier because they don't have to make as many changes as demand does. Second, when every company in an industry uses this method for setting prices, costs tend to be the same, which means there is less price competition. Third, a lot of people think that cost-plus price is better for both buyers and sellers. Sellers get a good return on their money, but they don't take advantage of buyers when there are a lot of them.

However, markup pricing continues to be widely favored for numerous reasons. Initially, sellers possess a higher level of certainty regarding costs compared to their level of certainty regarding demand. By correlating the price with the cost, merchants streamline the pricing process, eliminating the need for frequent revisions in response to variations in demand. Furthermore, when all companies within the industry adopt this pricing strategy, prices tend to converge and as a result, price-based competition is reduced to a minimum. Furthermore, a significant number of individuals believe that cost-plus pricing is more equitable for both purchasers and vendors. Sellers receive a justifiable profit on their investment without exploiting purchasers during periods of high demand.

3.5 Break-Even Analysis and Target Profit Pricing

Break-even pricing, often known as target profit pricing, is another pricing strategy that focuses on cost optimization. The company endeavors to ascertain the price at which it will reach the point of equilibrium or achieve the desired level of profit. General Motors in the United States utilizes this pricing strategy to ensure a profit margin of 15 to 20 percent on its investment in autos. Public utilities also employ this pricing mechanism, as they are obligated to generate a reasonable profit on their investments.

Break even pricing: fixing the price to cover the expenses of production, advertising, and promotion, or fixing the price to achieve a specific level of profit.

Target pricing employs the utilization of a break-even chart, which visually displays the anticipated total cost and total revenue at various levels of sales volume. The break-even point (BEP) is the point at which the total cost (TC) is equal to the total revenue (TR), resulting in neither profit nor loss. Figure 10.5 displays a break-even chart for the juice manufacturer that has been previously mentioned. The fixed costs amount to Rs. 300,000 irrespective of the sales volume. Total costs are the sum of fixed expenses and variable costs, and they increase as volume increases. The total revenue curve originates from zero and increases with each unit sold. The slope of the total income curve indicates a price of Rs. 20 per unit. The intersection point of the total income and total cost curves occurs at 30,000 units. This represents the point at which total revenue equals total costs, resulting in neither profit nor loss. In order to reach the break-even point, where total income equals total cost, the company needs to sell a minimum of 30,000 units at a price of Rs. 20 each. The break-even volume can be determined by applying the following formula:

$$\text{Brek Even Volume} = \frac{\text{Fixed Cost}}{\text{Price} - \text{Variable Cost}} = \frac{\text{Rs. }300{,}000}{\text{Rs. }20 - \text{Rs. }10} = 30{,}000$$

In order to achieve a desired level of profit, the corporation must exceed sales of 30,000 units, each priced at Rs. 20. Assuming that the juice producer has invested Rs. 1,000,000 in the firm, their objective is to set a price that will generate a 20 percent return, equivalent to Rs. 200,000. Therefore, it is necessary for it to achieve a minimum sales volume of 50,000 units, with each unit priced at Rs. 20. By setting a higher price, the corporation can attain its desired profit without having to sell a large number of units. However, it is possible that the market will not purchase this reduced quantity even if it is offered at a higher price. The outcome is heavily influenced by the price elasticity and the prices set by competitors.

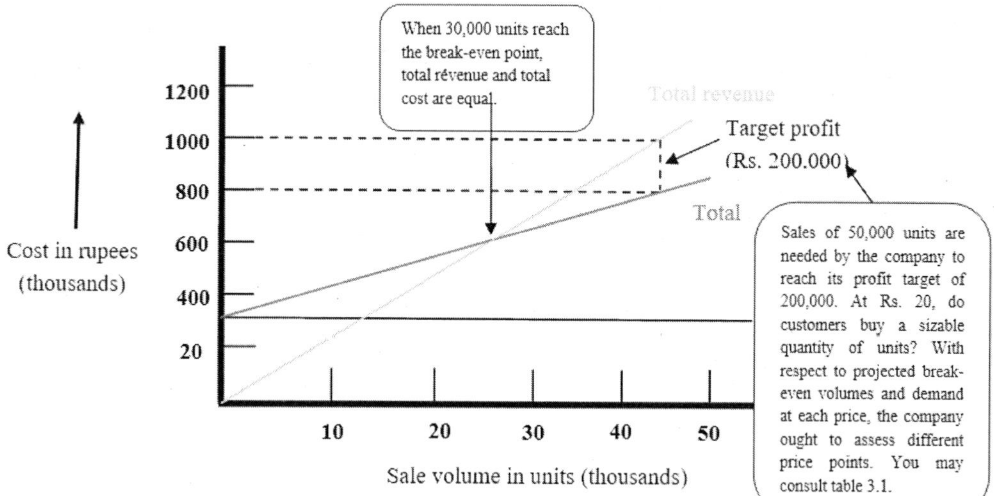

Figure 3.6 Break Even Chart for Setting Target Price

The manufacturer should calculate the break-even volumes, forecast the demand, and determine the profits for various price strategies. This information is located in Table 3.1. The break-even volume declines as the price lowers in column 2 of the table. There has been a decrease in the demand for juice due to an increase in its price (column 3). The manufacturer's profit margin is just Rs. 4 per pack, after deducting Rs. 10 in variable expenses from the selling price of Rs. 14. Therefore, a substantial volume of sales is required to reach the break-even point. The producer incurs losses as the demand for the product falls below the high break-even point, despite the low-price attracting clients. In order to reach the break-even point at a price of Rs. 22, the firm must generate a profit of Rs. 12 per juice pack and sell a total of 25,000 units. The negative profits are a result of insufficient consumer demand for juice packs at the current high pricing. According to the table, Rs. 18 generates the highest amount of money. None of the prices satisfy the manufacturer's profit targets of Rs. 200,000. In order to achieve this return, the manufacturer needs to reduce either fixed or variable expenses in order to decrease the break-even volume.

Price	Unit Demand required to Break Even	Expected Unit Demand at given price	Total revenue (1) x (3)	Total Cost	Profit (4) - (5)
1	2	3	4	5	6
Rs. 14	75,000	71,000	Rs. 994,000	Rs.1,010,000	-Rs. 16,000
Rs. 16	50,000	67,000	Rs. 1,072,000	Rs.970,000	Rs. 102,000
Rs. 18	37,500	60,000	Rs. 1,080,000	Rs.900,000	Rs. 180,000
Rs. 20	30,000	42,000	Rs. 840,000	Rs.720,000	Rs. 120,000
Rs. 22	25,000	23,000	Rs. 506,000	Rs.530,000	-Rs. 24,000

Table 3.1 Break Even Volume and Profit at Different Prices

3.5.1 Overall Marketing Strategy

Price is merely a single component within the company's wider marketing plan. Prior to determining the price, the corporation must establish its comprehensive marketing strategy for the product or service. If the organization has meticulously chosen its target market and positioning, then its marketing mix approach, encompassing price, will be quite uncomplicated. As an illustration, Honda has to set a high price for its Civic and Accord models in order to compete with other luxury-performance automobiles in the higher-income sector in India. These cars, similar to all other luxury sedans, had a price tag above Rs. 1,000,000. Contrarily, considering the rising need for affordable compact cars in India and the remarkable achievements of Maruti 800, Indica, and other economical and fuel-efficient vehicles, Honda intends to introduce the Honda Jazz model in the hatchback category, commonly known as the B segment, in India. This market positioning necessitates offering the vehicle at a competitive price. Therefore, decisions on market positioning have a significant impact on pricing strategy.

Pricing may significantly contribute to achieving organizational objectives across several levels. A company has the ability to establish prices in order to attract new clients or to maintain existing ones in a profitable manner. It has the ability to establish low prices in order to deter competition from entering the market, or to set prices at the same level as competitors in order to maintain market stability. It is beneficial to offer competitive pricing in order to maintain the loyalty and support of resellers, as well as to prevent government interference. Temporary price reductions might be implemented to generate enthusiasm for a brand. Alternatively, a product may be strategically priced to boost the sales of other products within the company's product portfolio.

Price is a single component of the marketing mix that a firm employs to accomplish its marketing goals. In order to create a cohesive and successful integrated marketing program, it is essential to align price decisions with product design, distribution, and promotion decisions. Pricing selections may be influenced by decisions made for other marketing mix components. For instance, if the decision is made to represent the product as high-performance quality, the seller will need to increase the price in order to compensate for the higher costs. Producers that anticipate that their resellers would provide assistance and actively market their items may need to incorporate higher reseller margins into their pricing.

Companies frequently strategize their product positioning based on pricing, and subsequently adjust other marketing mix elements to align with their desired prices. The price of a product is a critical aspect in determining its market position, competitiveness, and design. Target costing is a powerful strategic tool that many companies use to support price-positioning initiatives. Target costing involves a different approach from the traditional process of designing a new product, estimating its cost, and then evaluating its marketability based on that cost. Instead, it begins by establishing an optimal selling price that takes into account customer-value factors, and then aims to control expenses in order to achieve that price. When Tata Motors embarked on the design of the Tata Nano, it aimed to achieve a starting price point of Rs. 100,000 and operational efficiency comparable to that of the Maruti 800. Subsequently, it developed a fashionable and energetic compact vehicle at a price point that enabled it to deliver those desired qualities to its intended clientele.

Other companies downplay the importance of pricing and instead utilize various marketing strategies to establish positions based on factors other than price. Frequently, the optimal approach is not to set the lowest price, but rather to distinguish the marketing offer in order to justify a higher price. For instance, Bose, renowned for its state-of-the-art consumer electronics, incorporates additional worth into its items and sets exorbitant costs. As an illustration, a Bose speaker equipped with a sub-woofer from the most affordable category will have a price tag of Rs. 77,000. Is a comprehensive Bose audio system available? Frankly, you have no genuine interest in acquiring this information. However, the target buyers acknowledge the exceptional quality of Bose products and are willing to spend a premium price to obtain it.

Some marketers even position their products at high prices, incorporating the high prices as a component of the product's appeal. For instance, certain Titan timepieces are priced at more than Rs. 50,000 each. Fashion accessories are their designation, as opposed to mere timepieces.

Consequently, when establishing prices, vendors must take into account the comprehensive selling strategy and combination. The price will be significantly influenced by decisions regarding quality, promotion, and distribution if the product is positioned on nonprice factors. If price is a critical positioning factor, it will have a significant impact on the decisions made regarding the other elements of the marketing mix. However, marketers must bear in mind that consumers seldom make purchases solely based on price, even when price is prominently displayed. Instead, they focus on products that provide the most value in terms of the benefits they receive for the price they pay.

3.5.2 Organizational Considerations

Management is responsible for determining which employees within the organization are authorized to establish pricing. There are numerous methods by which companies manage pricing. Top management frequently establishes prices in small organizations, as opposed to the marketing or sales divisions. Divisional or product line administrators are typically responsible for pricing in large organizations. Salespeople may be permitted to negotiate with consumers within specific price ranges in industrial markets. Nevertheless, the pricing objectives and policies are established by the top management, which frequently authorizes the prices proposed by lower-level management or marketers.

In industries where pricing is a critical factor (such as airlines, aerospace, steel, oil companies and rail roads), companies frequently establish pricing divisions to either establish the most competitive prices or assist others in doing so. The marketing department or upper management is the source of authority for these departments. Sales managers, finance managers, production managers, and accountants are among the individuals who have an impact on pricing.

3.6 Break-Even Point

The break-even point means the level of output or sales at which no profit or loss is achieved. It indicates the position at which marginal profit or contribution is just sufficient to cover fixed overheads. In other words, a business is said to break-even when its income equals its expenditure. When production exceeds the "Break-even point", the business makes a profit and when it is below the "Break-even point", the business makes loss. This is shown in Figure 3.6.

The break-even point of any two variable situations is the point or the value at which they become equal as the result of a common variable.

There are two methods mentioned below to obtain break-even point:

- Mathematical method and
- Graphical method.

3.6.1 Mathematical Method

Let cost be the common variable in two situations 1 and 2, then cost equations will be

$$C1 = f1(x)\text{... a function of } (x) \quad \text{... (1)}$$

$$C2 = f2(x)\text{...another function of } (x) \quad \text{... (2)}$$

C1-may be as total cost, annual cost, cost per item or cost per day etc. for situation 1.

C2- same as C1 but application to situation 2.

x-variable effecting C1 and C2.

To solve for the value of x, let C1 = C2

i.e f1(x) = f2(x) ... (3)

Equation 3. can be solved for obtaining the value of x. The value of x making the cost equal in both the situations is called "Break Even Value". Below this value of x one situation will be economical while above it another situation will be economical.

Example 3.1. A 25 H.P. unit is required to drive a pump to remove water from a tunnel. The number of hours for which the power unit will run per year is dependent on weather conditions. The power unit is to be used for 4 years. For the supply of power following two plans are under consideration.

Plan I. This plan requires the construction of power line and purchase of electric motor at a total cost of Rs. 16,000. The salvage value of which is Rs. 4000 after 4 years of working. Cost of electricity per hour of operation is Rs. 6.80. Equipment being automatic, no attendant is needed. Maintenance is estimated to Rs. 2400 per annum.

Plan II. This plan needs a gasoline engine, which costs Rs. 11,000. The engine will be condemned at the end of 4 years. The cost of fuel and oil per hour of operation is estimated as Rs. 8.40. Hourly wages of operator is Rs. 2.00. Maintenance is estimated at Rs. 3 per hour of operation.

Solve by "Break-even point" theory, which of the plan will be economical?

Solution

Plan I. Let, N = No. of hours of operation per year

Then, total annual cost

$$= \{(16{,}000-4{,}000)/4\} + 6.8N + 2400$$

$$= 5400 + 6.8N \quad \text{... (1)}$$

Plan II. Similarly, total annual cost

$$= 11{,}000/4 + 8.4N + 2N + 3N \quad \text{... (2)}$$

There is one value of N for which the cost of Plans I and II will be equal. Hence N may be determined by equating equations (1) and 2).

$$5400 + 6.8\ N = 2750 + 13.4\ N$$

$$13.4\ N - 6.8\ N = 5400 - 2750$$

6.6 N = 2650

N = 401 hours.

For given conditions, annual costs of the two alternatives are calculated to be equal for 401 hours of usage per annum. If usage comes to be less than 401 hours per annum, selection of plan II is economical. For more than 401 hours, the selection for automatic equipment, i.e. plan I will be more economical

3.6.2 Graphical method

Although the break-even point may be calculated mathematically, but it is usually represented graphically because it enables manager to see more clearly the break-even point and the possibilities for profits and losses. By using these charts one can predict probable profits at various levels of output. A break-even chart given in Fig. 3.7 is used to determine break-even point and amount of profit or loss under changing conditions of output and costs. Sales or expenditure in rupees is represented on vertical axis, while output (either in quantity or in percentage capacity) is represented on horizontal axis. Line A rep- resents the "fixed cost", line B represents total cost or total expenses, while line C represents sales revenue and indicates income at various levels of output. The point where lines B and C intersect each other is "Break Even Point". The space between lines B and C to the right of the "Break Even Point" represents potential profit, whereas to the left of the "Break Even Point" potential loss. The amount of loss or profit can be measured on vertical scale.

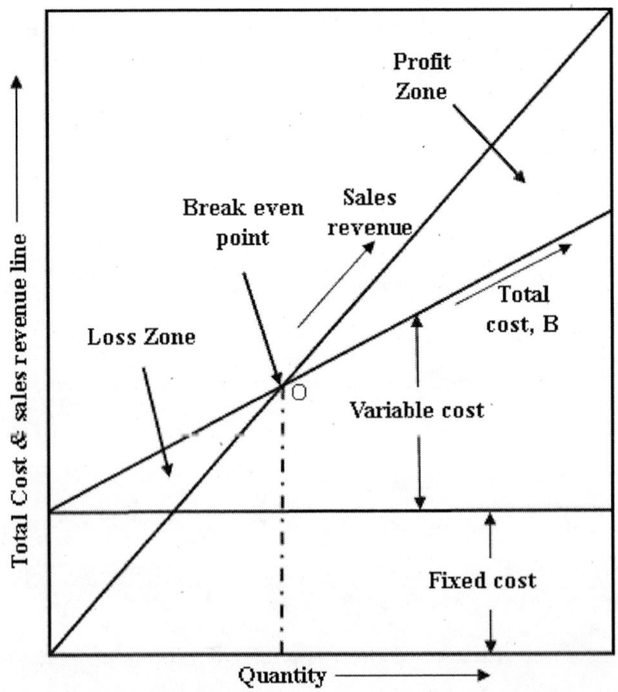

Figure 3.7 Illustrate the concept of break-even point and amount of profit or loss under varying conditions of output and costs.

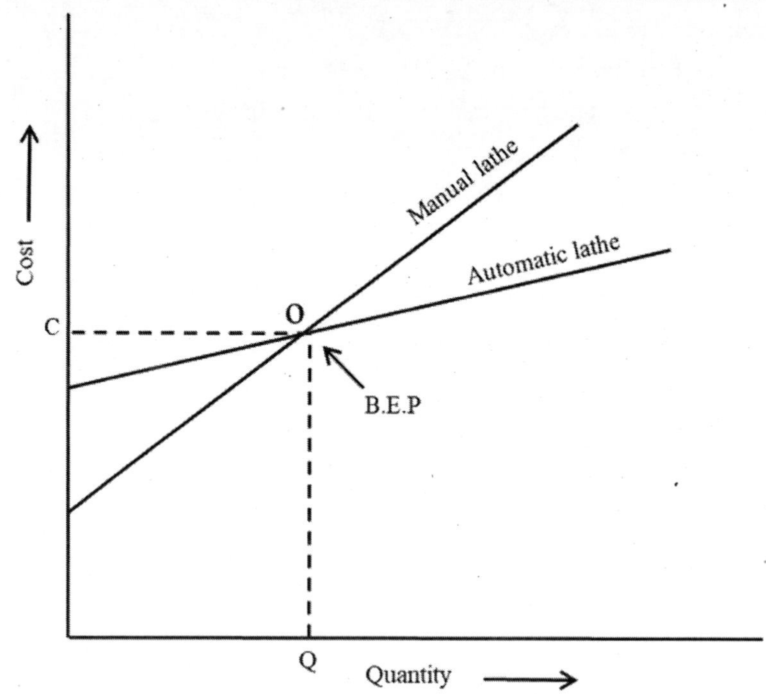

Figure 3.8 Break Even Point Chart to Decide Production Quantity

This method can be applied to various management problems. For example, suppose a manager wants to replace an old lathe machine being used for manufacturing screws by automatic screw machine. Then he must first know whether it will be profitable or not, for which he must adopt break-even point theory and construct the chart as explained in Fig. 3.8. The figure shows that for a production less than Q, it must not be changed whereas for production more than Q, automatic machine or new machine will be economical or in other words below Q manual lathe is cheaper; to beyond Q, automatic machine is cheaper. This break- even point is also known as "cut even point".

3.7 Some Important Definitions

1. **Angle of incidence**
   It is the angle at which income line or sales line cuts the total cost line. If the angle is large, it is an indication that profits are being made at a high rate, on the other hand, if the angle is small, it indicates that less profits are being made and are achieved under less favorable conditions.
2. **Margin of safety: -**
   It is the output at full capacity minus the output at "Break Even Point". It is expressed as percentage of output at full capacity. If the margin of safety is small, a small drop in production capacity will reduce the profit greatly. It can also be expressed as:

Margin of safety = ((Sales at full capacity-Sales at B.E.P)/ (Sales at full capacity)) ×100

3. **Contribution: -**

   It is the difference between sales and variable cost (marginal cost). It is also called as Marginal Profit or Gross Marginal. The marginal profit provides the contribution towards fixed cost and profit.

3.7.1 Break-Even Point Calculations

Let S = Sales price,

V=Variable cost

F = Fixed cost and P = Profit

Now $$S = F + V + P$$

Or $$S - V = F + P \quad \ldots(i)$$

At break-even point, P=0.

$$S - V = F \quad \ldots(ii)$$

Multiplying both sides of Eq. (ii) by S.

$$\therefore \quad S(S-V) = F \times S$$

or $$S = (F \times S)/((S-V)) = F/(((S-V)/S)) = (\text{Fixed Cost})/(\text{Contribution per unit})$$

and Sales at B.E.P. $$= \text{Rs. } (F \times S)/((S-V)) \quad \ldots(iii)$$

and

No. of units at B.E.P.= (Fixed cost)/(Contribution cost)=(Fixed cost)/(Marginal cost) ....(iv)

3.7.2 Break Even Analysis (Cost Analysis)

This is also known as cost analysis. Break-even analysis is concerned with finding the point at which revenues and costs are exactly equal. This point is known as BREAK-EVEN-POINT. Thus, this is a volume of output at which neither a profit is made nor a loss is incurred. Therefore, production or sale must not be allowed to fall beyond this point.

This analysis can be carried out either algebraically or graphically.

3.7.3 Break-even chart

A breakeven chart is a graphical depiction of the relationship between costs and revenue at a given time, and determines the break-even-point and profit potential under varying conditions of output and cost.

3.7.4 Functions of Breakeven Chart

- To represent economical position of production on graph.
- To for tell likely profits or losses at various levels of output.
- To help the management to decide the production level.
- To indicate margin of safety.

3.7.5 Determination of the Break-Even-Point

It may be determined in terms of physical units or in money terms i.e. sales value in rupees:

3.7.5.1 Break-Even-Point in terms of physical units

Break even volume is the number of units of a product which must be sold to earn enough revenue just to cover all expenses. The break-even-point (BEP) is reached when sufficient number of units have been sold so that the total contribution margin of the units sold is equal to the fixed costs.

B.E.P= (Fixed Cost)/(Selling price-Variable cost per unit)

3.7.5.2 Break-Even-Point in terms of Sales Value

Multi product firms are not in a position to measure the BEP in terms of any common unit of product. In these firms it is convenient to determine their BEP in terms of total rupee sales. In this case BEP would be the point where the contribution margin (Sales value-Variable costs) would be equal to fixed costs contribution margin is expressed as a ratio to sales.

B.E.P= (Fixed Cost)/(Contribution ratio)

Where, Contribution ratio= (Sales value-Variable costs)/(Sales value)

- **Margin of Safety.** This is shown on the chart by the distance between B.E.P. and the output being produced. It shows that if this distance is short then a small decrease in output or sales will reduce the profit greatly. If the distance is long, it means the business could still be making profit after a great reduction in output.
- **Angle of Incidence**. By cutting sales line on to a total cost line an angle known as "Angle of Incidence" is formed. Chart shows that if the angle is large, it is an indication of large profits and if it is small it shows that profit are being earned under less favorable conditions.

3.7.5.3 Position of Break-Even Point

If B.E.P. is over to the left of the chart with a large angle of incidence it shows that output can be raised considerably. If B.E.P. is over to the right of the chart, the margin of safety is low, which means:

- the fixed overheads are too great for the amount of sales being done, and
- the fixed and variable costs are high while the profit is small.

If the production volume is below B.E.P., the company will be running in loss and beyond it profit can be had. Break even-point may be determined in terms of physical units or in money terms.

- B.E.P. in terms of physical units. This is convenient for the single product firm. It represents the number of units of a product which must be sold to earn enough revenue just to cover all expenses.
- B.E.P. in terms of sales value. Multi-product firms are not in a position to measure the B.E.P. in terms of any common unit of product. In these firms it is convenient to determine this B.E.P. in terms of total rupee sales.

3.7.6 Assumptions underlying Break-Even Analysis

1. All the costs are either perfectly variable or absolutely fixed over the entire range of production.
2. All revenue is perfectly variable with the physical volume of production.
3. The volume of sales and the volume of production are equal.
4. In case of multi product firms, the product mix should be stable.

3.7.7 Applications

Break even analysis not only highlights the areas of economic strength and weaknesses in the firm but also helps in finding out the ways which can enhance its profitability. With the help of this analysis management of a production firm can take decisions related to the following:

1. Safety margin. It decides the extent to which the firm can afford to decline in sales, before it starts incurring losses.
2. Volume needed to attain target profit.
3. Change in price, and its effect.
4. Whether to expand production capacity or not.
5. Whether to add a new product or drop production of any product.
6. Whether to make or buy.
7. Selection of production machinery so as to get maximum profit for a particular volume of the product out of the available machineries.
8. Improving profit performance by
    - **I.** increasing the volume of sales, and or
    - **II.** increasing the selling prices, and or
    - **III.** reducing the variable expenses per unit, and or
    - **IV.** reducing the fixed costs.

3.7.8 Limitations.

1. Since Break Even Analysis is based on accounting data therefore, it can be sound and useful only if the firm in question maintain a good accounting system.
2. It is based on the assumptions of given relationships between costs and revenues, on one hand, and input on the other.
3. Cost data of the past period may not hold good for the current period.
4. Selling costs may not remain constant.
5. The cost revenue-Volume relationship is linear. But this is realistic only over narrow ranges of output.
6. Break even analysis is not an effective tool for long range use and its use should be restored to the short run only.

Example 3.2 Fixed costs in a factory is Rs. 10,000 per year, the variable costs are Rs. 2.00 per unit and the selling price is Rs. 4.00 per unit, calculate B.E.P.

Solution. B.E. P = (Fixed Cost)/(Contribution margin per unit) = 10,000/(4 – 2) = 500 Unit.

Check Sales = 5000 x 4 = Rs. 20,000

Cost of goods sold = Variable cost + Fixed costs = 5000 x 2 + 10,000 = Rs. 20,000

∴ Net profit = Nil

Example 3.3 If in example 1 sales are 8000 units, calculate safety margin. If desired profit is Rs. 6000, calculate target sales volume.

Solution. Safety margin = ((Sales-BEP))/Sales×100

= (8000-500)/8000×100=3/8×100=37.5% Ans.

Target sales volume = (10,000-6,000)/(80004-2)=100Units.Ans.

3.7.9 Safety Margin

It refers the extent to which the concern can afford to decline in sales up to the break-even point. It can be represented by the percentage ratio of sales over break-even point volume to the sales volume.

Safety margin=(Sales volume-Sales at B.E.P)/(Contribution per unit)×100

3.7.10 Quantity needed to have desired profit

This is the most common application of break-even analysis and the desired quantity is given

By Target quantity=(Fixed cost-Target profit)/(Contribution per unit)

3.7.11 The effect of change in price

Sometimes management is required to consider whether to reduce price of a product or not. Management has to consider the following facts in this regard:

If the sales price is reduced then contribution margin will reduce and this will increase the sales volume. But this increased volume may or may not receive increase in demand because it depends on elasticity of demand.

The change in break-even point due to rate variation in sales price is drawn below (Fig 3.9). At different price levels total revenue line can be plotted and the chart shows that as the price of a product is reduced the break-even point shifts towards right side and vice versa. From the chart, one can then estimate the volume that will be sold and profit at each price level.

From the chart

At B.E.P., Q = (F + P)/(S - V)

Q1 = (F + P)/(S_1 - V)

Here

S1 = New changed sales price.

Q1 = New changed B.E.P. in units

3.7.12 Effects of change in costs

There are two types of costs:

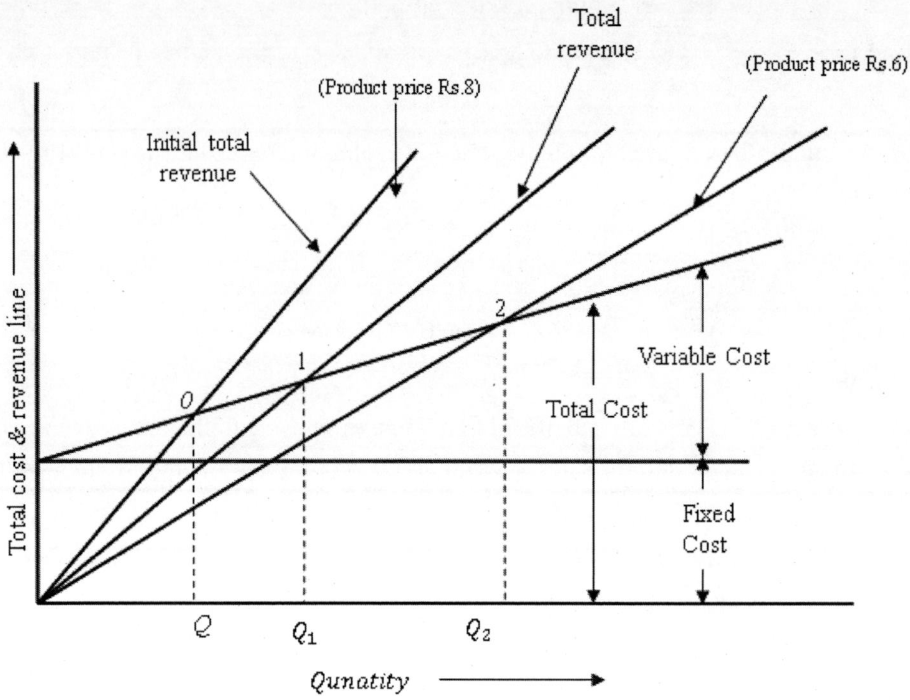

Figure 3.9 Effect on Break Even Point Due to Change in Price

1. Variable cost and
2. Fixed cost.

3.7.12.1 Variable cost change

An increase in variable cost leads to reduction in the contribution margin. This increase in variable cost will reduce the profit. Therefore, what should be the new price to maintain the present profit without any change in sales volume?

As there is reduction in contribution margin, it will cause increase in sales volume. Therefore, what should be the sales volume to maintain the present profit without change in the price (Fig. 3.9)

Where Q1 = New Break-even point and

V1 = New Variable cost.

If sales volume is same, then Q = Q1

$$\therefore \quad S_1 - V_1 = S - V$$

$$S_1 = S + (V_1 - V) \qquad ...(1)$$

where S_1 is the new selling price to maintain present profit.

Due to increase in the variable cost, Break-even point has shifted from point Q to point Q1 refer figure 3.10. Thus, higher quantity is to be sold to maintain the present profit assuming no change in the price of the product.

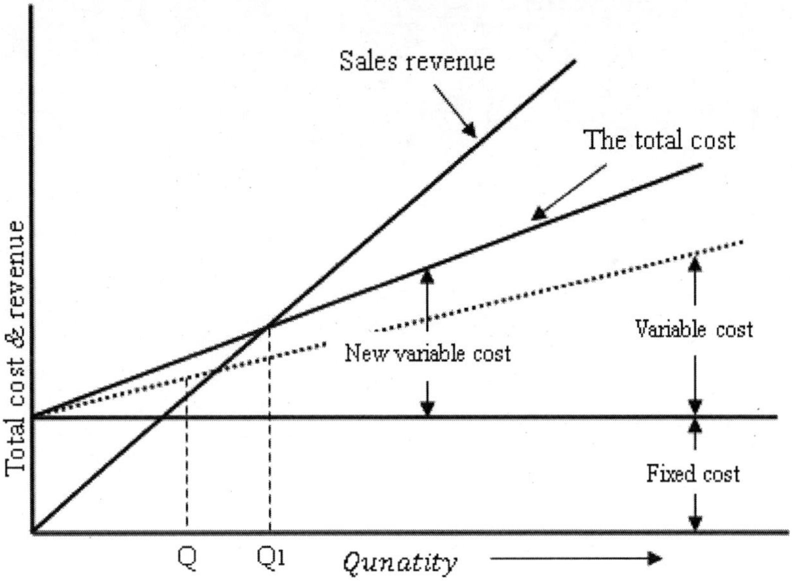

Figure 3.10 Effect on Break Even Point without Changes in Price

Change in fixed cost. The fixed cost may change due to an increase in executive salaries, taxes, insurances and building rent etc. This in turn will bring change in the break-even point, assuming no change in product price.

Here, $Q=(F=P)/(S-V)$ (1)

$Q_1= (F=P)/(S-V)$ (2)

Solving Equations (1) and (2),

we have $Q_1=Q+ (F_1-F)/(S- V)$ (3)

And if Q is same and there is change in sales price, then

$S_1=S+ (F_1-F)/Q$ (4)

Here, F_1=New fixed cost

This is shown graphically in Fig. 3.11 The Break-even point has shifted from point Q to Q_1. If there is no change in price and volume of profit is maintained, if there is no change in the sales volume, provided profit is maintained.

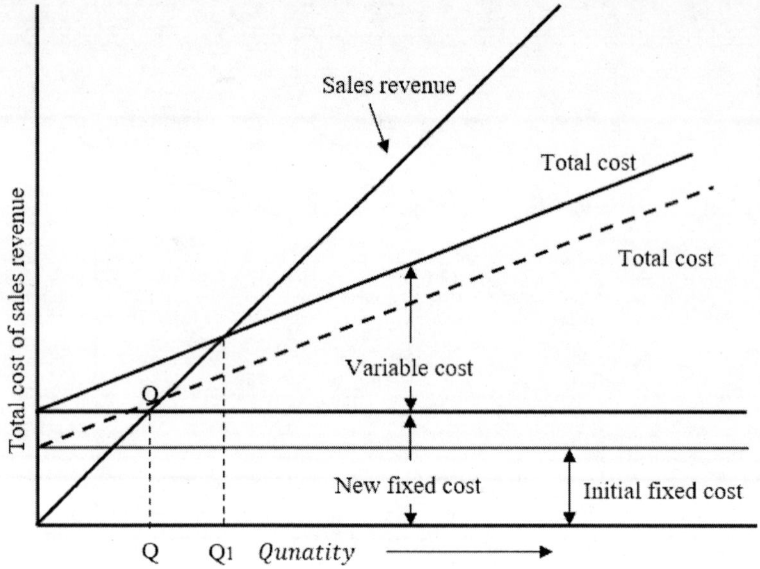

Figure 3.11 Effect on Break Even Point with Change in Fixed Cost

3.7.13 Whether to accept an order or not

If extra capacity to produce a product is variable then sometimes management has to decide whether to accept another order at a price with less than the initial selling price. For example, a plant is operating on its 50% of full capacity. A buyer offered to buy 4000 units at Rs. 5 per unit while the average cost is Rs. 6. Now the problem is whether to accept this order or not. The other details of the product are:

- Total capacity = 20,000 units
- Present output = 10,000 units
- Direct material cost = Rs. 10,000
- Direct labour cost = Rs. 10,000
- Fixed costs = Rs. 40,000

In this question, one can find that additional cost of making these 4000 units is Rs. 4000 for material cost and Rs. 4000 for labour cost (total Rs. 8000) and the revenue from sales would be

Rs. 4000 x 5 = Rs. 20,000

Thus, the concern will save,

Rs. 20,000 Rs. 8,000 Rs. 12,000

So, it would be profitable to accept anything more than Rs. 2 per unit which is the direct cost per unit. There is only danger that, if other consumers come to know that this product is being sold at a lower price, they may also then demand supply at the lower price or otherwise they may not be further interested to buy from this concern. Therefore, in such cases acceptance should be kept secret.

QR For Additional Learning About Break Even Analysis

3.8 Introduction of Aggregate operation planning

Manufacturing things is hard. Some companies only make a few items, while others make a lot of them. But each one uses a different set of methods, tools, machines, materials, work skills, and materials. For a business to make money, it needs to coordinate all of these things so that it can make the right goods at the right time, at the best quality, and for the least amount of money. It is hard to solve, so you need a good method for planning and keeping track of things. These four things must be true of any good planning system:

1. What are we going to sell or make for people?
2. What do you need to make it?
3. What do we have?
4. What do we need?

It comes down to priorities and resources.

Priority has to do with what goods are needed, how many are needed, and when they need to be bought. The objectives are set by the market. Manufacturing is in charge of making plans to meet market needs if they can.

Capacity is the number of things and services that a factory can make. Actually, it depends on what the company has available, like its tools, workers, money, and the materials it can get from sellers. In the short term, capacity is the amount of work that people and machines can do in a certain amount of time. Figure 3.12 shows a graph of the link that should exist between priority and capacity.

Long-term and short-term, manufacturing needs to make plans to match what the market wants with what it can and can't do. The plans must go back several years in order to make long-term decisions, like whether to build new plants or buy new tools. Time frames of days or weeks will be used to plan work for the next few weeks. The next part talks about this planned hierarchy, which goes from long-term to short-term.

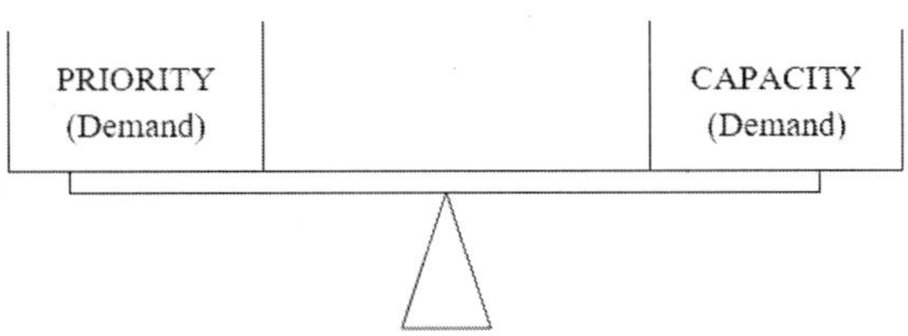

Figure 3.12 Relationship of Priority and Capacity

3.8.1 Manufacturing planning and control system

The major parts of the Manufacturing planning and control system are the following:

- Strategic business plan (a business plan based on the strategy).
- Production plan (sales and operations plan).
- Master production schedule.
- Material requirements plan.
- Purchasing and production activity control.

Each level differs in its goal, duration, and amount of intricacy. As the process transitions from strategic planning to production activity control, the objective shifts from providing overall guidance to doing particular and detailed planning. The time frame narrows down from years to days, and the level of specificity grows from broad categories to individual components and workstations.

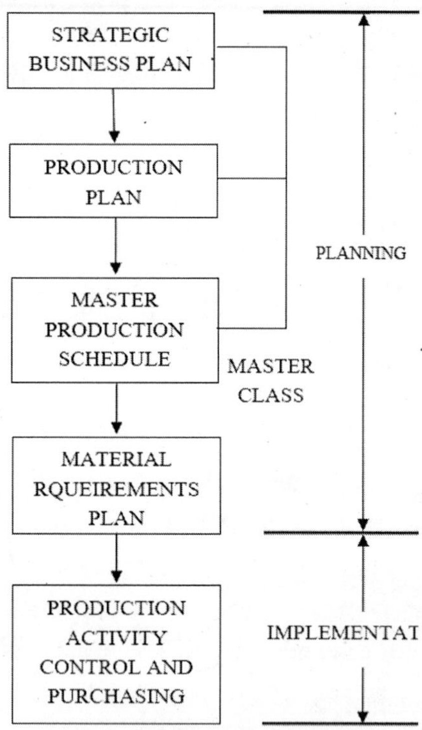

Figure 3.13 Manufacturing Planning and Control System

Due to variations in time span and intended aims, each level exhibits differences in the following aspects:

- The plan's objective.
- The planning horizon refers to the period of time, starting from the present and extending into the future, for which a plan is developed.
- The level of detail refers to the extent of information needed about the products for the strategy.
- The planning cycle refers to the frequency at which the plan is evaluated.

At each level, three questions must be answered:

1. What are the production priorities in terms of quantity and timing?
2. What is the current capacity of resources that we have available?
3. How might discrepancies between priorities and capacity be reconciled?

Figure 3.13 depicts the hierarchical structure of the planning process. The initial four stages are categorized as planning levels. The outcome of the plans is the authorization to procure or produce the necessary items. The ultimate stage is reached when the plans are implemented through the process of procurement and the control of manufacturing activities.

The subsequent sections will analyze each planning level based on its objective, time frame, amount of specificity, and planning frequency.

3.8.2 The Strategic Plan

A firm's strategic plan is a concise declaration of the primary goals and objectives that the company aims to accomplish within a timeframe of 2 to 10 years or beyond. A strategic statement outlines the overall objectives and goals of the company, including the specific business-product lines, target markets, and other areas of focus that the company intends to pursue in the future. The plan outlines the company's strategic objectives and serves as a commitment to implement specific measures aimed at achieving growth, acquiring customers, identifying markets, and enhancing competitive and financial performance. The process relies on extended-term predictions and involves the involvement of various departments within the company, such as marketing, finance, production, and engineering, as well as other key activities. The plan facilitates guidance and synchronization among the marketing, production, finance, engineering, and other functional plans.

The Marketing and Sales departments are tasked with evaluating the market and determining the company's course of action, including which markets to target, what items to offer, the desired degree of customer service, price, promotion techniques, and other related factors.

The finance department is accountable for determining the origins and allocation of funds accessible to the company, cash inflows and outflows, profitability, return on investment, and financial plans.

Production must meet the requirements of the market. It achieves this by optimizing the utilization of plants, machinery, equipment, manpower, and materials.

Engineering is accountable for conducting research, carrying out development activities, and designing novel items or making alterations to current ones. The field of engineering necessitates collaboration with marketing and production departments to develop product designs that are marketable and cost-effective to manufacture.

Senior management is responsible for developing the strategic plan. The strategic plan utilizes data from several departments such as marketing, finance, and production to establish a framework that outlines the aims and objectives for future planning by departments including marketing, finance, engineering, and production.

Setting mission statements and goals as part of the planning process is something that some companies do in a unique way. For the visions, tactical work plans are made so that everyone in the company can move in a planned way toward the overall goal. The method was created by Japanese companies and is often called **Hoshin Planning** in Japanese. These are the basic steps:

1. Making a plan for what you want to learn or do better.
2. Setting smaller goals.
3. Telling everyone in the company about the plan.
4. Keeping track of your progress.
5. Looking at the data from the tests and fixing things as needed.
6. Doing it again if needed.

Emerging corporate trends might occasionally influence the formulation and execution of strategic plans. One of these concerns is sustainability, which refers to the ability to maintain operations over the long term. The concept of sustainability has mostly emerged from concerns over pollution control, environmental preservation, and the promotion of social responsibility. This involves implementing firm policies that foster a constructive relationship with society and strive to achieve a balance between economic growth and environmental conservation. The United Nations Global Compact has highlighted the increasing significance of corporate social responsibility on a global scale. The compact acknowledges that business plays a significant role in the process of globalization. It outlines ten principles that businesses should follow in their strategies and operations to ensure the protection of human rights, fair treatment of workers, environmental sustainability, and the prevention of corruption.

Sustainability is predicated on the principles of **waste reduction and increased production efficiency**, resulting in the utilization of fewer resources, generation of less waste, and reduced costs. Illustrations of waste reduction could encompass a decrease in the necessity for packaging, which is frequently discarded, as well as the utilization of resources to generate reusable outputs. Occasionally, this process is referred to as remanufacturing or reverse logistics. Companies may build a structured supply chain to efficiently collect a used product for the purpose of disposing of it, extracting materials from it, or repurposing it. This is occasionally known as a reverse supply chain.

Another recent advancement that has an impact on strategic planning is the use of **risk management**. Frequently, individuals perceive danger as a detrimental matter, and indeed, that viewpoint is somewhat accurate. Risks stemming from system failures, human errors, or external events can lead to financial losses, decreased productivity, legal complications, and hinder the proper execution of the strategic strategy. Positive risks, known as opportunities, might also exist in this setting. Risk management entails the establishment of systems and metrics to promptly identify risks and implement strategic processes to mitigate the consequences of negative risks and capitalize on positive risks (opportunities).

Efficient strategic planning relies on acquiring suitable metrics and feedback regarding the effectiveness of the tactics in relation to the organization's overarching strategic goals. These metrics, encompassing both monetary and non-monetary aspects, are commonly known as **Key Performance Indicators (KPIs).** It is crucial to ensure that no individual metric or group of measurements contradicts others, and that they collectively provide a comprehensive assessment of the company's progress towards its strategic plan and sustainability objectives, encompassing financial, societal, and environmental goals. KPIs that exhibit balance are commonly known as a balanced scorecard, and specific methodologies have been devised to create and oversee balanced scorecards. The scorecard aims to achieve a balance between measurements related to company processes, financial performance, customer satisfaction, and learning and development. These views and measures are produced as part of the strategic planning process.

3.8.3 The Strategic Business Plan (Business Plan)

he strategic plan is frequently restated in financial terms after it has been established, which includes a projected balance sheet, projected income statement, and projected revenues. This financial-based strategy is frequently referred to as the business plan or, on occasion, the strategic business plan. In order to accomplish the strategic business plan's objectives, each department develops its own plan. The strategic business plan and these plans will be in close coordination. This relationship is illustrated in Figure 3.14.

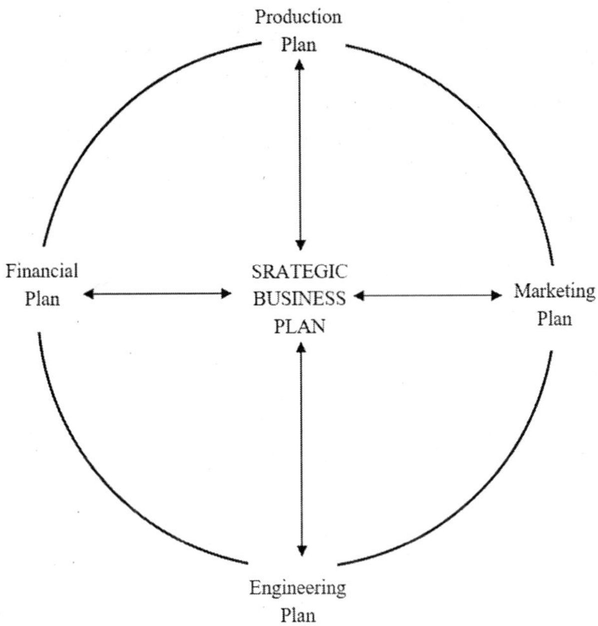

Figure 3.14 Business Plan

The strategic business plan lacks sufficient granularity. This pertains to the whole market and production needs, encompassing the entire market for large product groups or families, rather than focusing on the sales of individual items. The value is typically expressed in terms of currency, specifically dollars, rather than in terms of individual units.

Authors Comment: Strategic business plans are usually reviewed every six months to a year.

3.8.4 The Production Plan

Production management focuses on the following aspects in accordance with the objectives outlined in the strategic business plan:

- The production volumes required for each product group in each time.
- The optimal inventory levels.
- The required resources, including equipment, personnel, and material, for each phase.
- The accessibility of the required resources.

The production strategy lacks sufficient granularity. For instance, in the case of a company engaged in the manufacturing of children's bicycles, tricycles, and scooters, each available in multiple models and with varying features, the production plan will outline the primary product categories or families, namely bicycles, tricycles, and scooters. The production plan is commonly known as **the aggregate production plan** since it typically involves grouping product families or product groupings rather than individual goods.

Production planners are responsible for creating a strategy to meet the demand of the market using the company's existing resources. This will entail assessing the necessary resources to fulfill market demand, evaluating them against the available resources, and formulating a strategy to reconcile the requirements and availability.

Capacity management is the process of assessing the necessary resources and comparing them to the resources that are currently available. This evaluation occurs at every stage of planning and is the main objective of capacity management. In order to plan effectively, it is crucial to maintain a harmonious equilibrium between priority and capability.

In addition to the market and financial plans, the production plan focuses on executing the strategic business strategy. The typical time frame for the planning horizon is from 6 to 18 months and is regularly assessed on a monthly or quarterly basis.

3.8.5 The Master Production Schedule

The master production schedule (MPS) is a strategic plan that outlines the production schedule for each particular end item. The production plan is analyzed to display the quantity of each end item to be produced for each period. For instance, it could indicate that a total of 200 Model A23 scooters will be manufactured on a weekly basis. The inputs to the MPS include the production plan, the forecast for each specific end item, sales orders, stocks, and current capacity.

The level of specificity for the Master Production Schedule (MPS) surpasses that of the production plan. While the production plan was organized around groups of products, such as tricycles, the master production schedule is created for each particular end item, such as each model of tricycle. The planning horizon typically spans from 3 to 18 months, with its duration mostly determined by the lead times for purchasing and manufacturing. Master scheduling refers to the procedure of creating a master production schedule. The term "master production schedule" represents the final outcome of this process. Typically, the plans undergo weekly or monthly reviews and modifications.

3.8.6 The Material Requirements Plan

The material requirements plan (MRP) is a strategic plan that outlines the necessary components for production and procurement in order to fulfill the items specified in the master production schedule. The document displays the required quantities and the intended timing for manufacturing or utilization. The procurement and production activity control utilize the Material Requirements Planning (MRP) system to carry out the acquisition or production of particular items.

The level of specificity of MRP is substantial. The material requirements plan determines the specific timing for the procurement of components and pieces required for the production of each final product.

The planning horizon is equal to or more than the sum of the purchasing and manufacturing lead periods. Similar to the master production schedule, it typically spans a duration of 3 to 18 months.

3.8.7 Purchasing and Production Activity Control

Purchasing and production activity control (PAC) is the stage in the production planning and control system when the execution and control of activities related to purchasing and production take place. The purchasing department is responsible for the procurement and management of raw materials that are required for the factory's operations. The PAC is accountable for strategizing and managing the progression of tasks within the factory.

The planning horizon is rather brief, typically ranging from one day to one month. The level of granularity is substantial as it pertains to specific elements, workstations, and orders. Plans undergo daily assessment and revision.

Figure 3.15 illustrates the correlation between different planning tools, planning timeframes, and the level of detail.

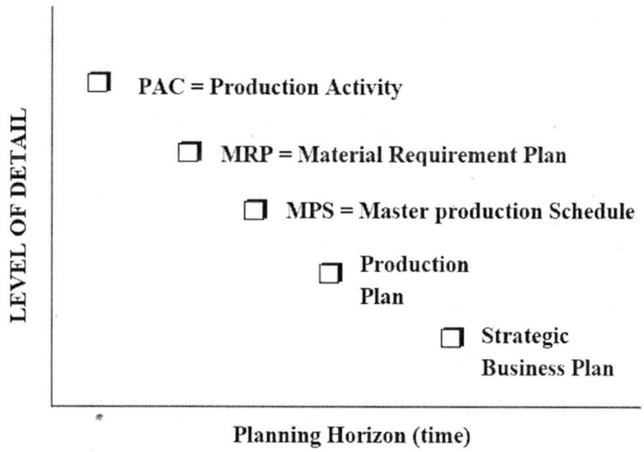

Figure 3.15 Level of Detail Versus Planning Horizon

3.9 Capacity Management

At every stage of the manufacturing planning and control system, the priority plan needs to be evaluated in relation to the resources and capability of the manufacturing system. It is important to comprehend that the fundamental procedure in capacity management involves determining the necessary capacity for manufacturing according to the priority plan and devising strategies to ensure that capacity is accessible. Without completing this task, it is impossible to have a viable and effective manufacturing plan. If the necessary capacity cannot be provided at the required time, then it is necessary to alter the plans.

At every level of the factory planning and control system, it is necessary to determine the required capacity, compare it to the available capacity, and make adjustments or change plans accordingly.

Manufacturing operations can undergo changes in machinery, equipment, and plants over a span of multiple years. These time intervals allow for various modifications to be made, such as adjusting the number of shifts, implementing overtime labor, subcontracting tasks, and more. However, it is not feasible to make significant modifications between the time intervals between production planning and production activity control.

3.10 Sales and Operations Planning

The strategic business plan consolidates the plans of all organizational departments and is often revised on a yearly basis. Nevertheless, it is imperative to regularly revise these plans to incorporate the most recent projections, as well as consider the prevailing market and economic circumstances. Sales and operations planning (S&OP), often known as SOP, is a systematic procedure for consistently updating the strategic business plan and synchronizing the plans of different departments. A Standard Operating Procedure (SOP) is a comprehensive corporate strategy that encompasses various departments like sales and marketing, product development, operations, and senior management. Operations symbolize the provision of goods or services, whereas marketing symbolizes the desire or need for those goods or services. The Standard Operating Procedure (SOP) serves as the platform for the creation of the production plan.

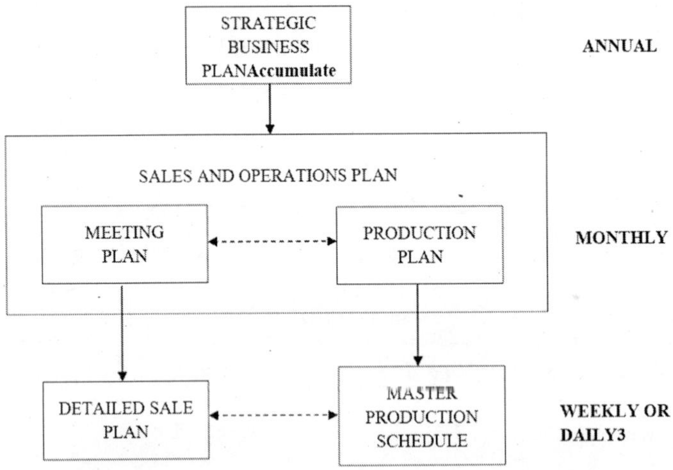

Figure 3.16 Sales and Operations Planning

While the strategic plan and strategic business plan are typically revised on a yearly basis, sales and operations planning is a fluid process that involves constant updates to the company's plans, usually on a monthly basis at the very least. The process commences with the sales and marketing departments, who analyze the current demand in relation to the sales plan, evaluate the market potential, and predict future demand. The amended marketing plan is subsequently sent to the manufacturing, engineering, and finance departments, who then make necessary modifications to their respective plans in order to align with the updated marketing strategy. If these departments determine that they are unable to adapt the new marketing plan, then it is necessary to make adjustments to the marketing plan. The strategic business plan is consistently updated throughout the year, ensuring that the actions of different departments are synchronized. Figure 3.16 depicts the correlation between the strategic business plan and the sales and operations strategy.

Sales and operations planning encompasses the marketing, manufacturing, engineering, and finance plans, and operates within a medium-range time frame. Sales and operations planning offers numerous advantages:

- It offers a method for revising the strategic plan and strategic business plan in response to changing conditions.
- It offers a method for effectively handling and overseeing change. Instead than responding to market conditions or economic changes after they occur, the Standard Operating Procedure (SOP) compels management to regularly assess the economy, enabling them to proactively strategize and adapt.
- It guarantees that the plans of different departments are both feasible and aligned, while also contributing to the overall business strategy.
- It offers a pragmatic strategy that can effectively accomplish the company's goals.
- It enables more efficient control over production, inventory, and backlog (i.e., outstanding client orders).

Effective sales and operations planning is a high-level planning process that requires executives to make crucial decisions that may involve major trade-offs between different roles or departments inside the firm. By adopting this technique, executives responsible for planning may guarantee the optimal alignment of volume and mix, ensuring a balanced supply and demand.

It is important to emphasize that the sales and operations plan does not directly schedule production. Rather, its main objective is to create a comprehensive strategy for the allocation of corporate resources. These resources encompass not only production resources but also people resources, sales resources, financial resources, and essentially all other operations inside the organization. The sales and operations plan should align with the vision and strategies outlined in the strategic plan. It should serve as a comprehensive and unified strategy that covers all functional areas and outlines the approach for managing the complete firm. Tom Wallace, the author of Sales and Operations Planning, outlines a five-step process for the development of the S&OP, which is summarized below:

1. **Data collection**. Actual sales from the previous month, current inventory levels, financial data, marketing/sales data, and so forth.
2. **Demand forecasting**. Establishing management forecasts by utilizing data and other inputs from all relevant sources. Statistical forecasts may be generated; however, they should be assessed in conjunction with other inputs. These inputs may encompass competitive movements, economic conditions, price changes, and new product introductions. Chapter 8 of

this book provides a more comprehensive explanation of demand planning and demand management.

3. **The planning of supplies**. Comparison of capacity constraints with demand forecasts.
4. **A pre-S&OP meeting**. This meeting should be utilized to establish an agenda for the top manager S&OP meeting, resolve any discrepancies, and develop action recommendations.
5. **The executive meeting.** The company's ultimate decisions regarding the course of action are determined by the data, recommendations, and the extent to which the plan aligns with the strategic plan and strategic business plan.

3.11 Manufacturing Resource Planning

It is probable that the manufacturing planning and control system will be computer-based due to the abundance of data and the numerous calculations required. The time and labor required to perform calculations manually are substantial, necessitating that a company make compromises if a computer is not utilized. The company may be compelled to increase inventory and extend lead times in order to compensate for the incapacity to promptly schedule the necessary items and their respective dates, rather than utilizing the planning system to schedule.

The system is designed to function as a completely integrated planning and control system that operates from the top down and receives feedback from the bottom up. Strategic business planning is the process of combining the plans and activities of marketing, finance, and production to develop plans that are designed to attain the company's overall objectives. Subsequently, the objectives of the production and strategic business plans, as well as the company as a whole, are pursued through the implementation of master production scheduling, material requirements planning, production activity control, and procuring. If capacity constraints necessitate modifications to priority plans at any of the planning levels, these modifications should be reflected in the subsequent levels. Subsequently, feedback must be implemented throughout the system.

The marketing, finance, and production plans are all included in the strategic business plan. Marketers must concur that their strategies are feasible and feasible. Finance must concur that the plans are financially advantageous, and production must concur that they can satisfy the necessary demand. The primary game plan for all departments in the company is the manufacturing planning and control system, as it is described in this document. A manufacturing resource planning (MRP) II system is the name given to this comprehensive planning and control system. The term "MRP II" is employed to differentiate between the material requirements plan (MRP) and the manufacturing resource plan (MRP).

Marketing and production are coordinated by MRP II. The production plan is a comprehensive and feasible strategy that has been agreed upon by finance, marketing, and production. Marketing and production must collaborate on a daily and weekly basis to modify the plan in response to evolving circumstances. It may be necessary to modify order sizes, terminate orders, and adjust delivery dates. This type of modification is implemented through the master production schedule. Marketing managers and production managers may modify master production schedules to accommodate fluctuations in anticipated demand. The production plan may be modified by senior management to account for changes in demand or resources. Nevertheless, they are all operational via the MRP II system. It serves as the means by which the company's marketing, finance, production, and other departments collaborate. MRP II is a technique that enables the efficient allocation of all resources within a manufacturing organization.

Note: Figure 3.17 illustrates the feedback circuits in the MRP II system

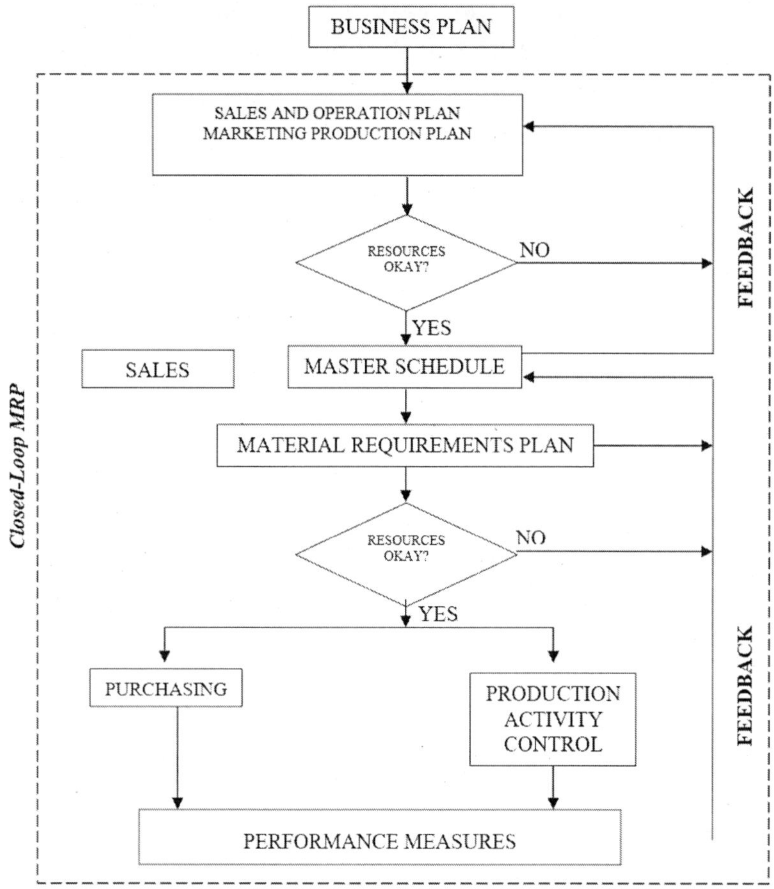

Figure 3.17 Manufacturing Resources Plan (MRP II)

3.12 Enterprise Resource Planning

MRP systems tended to capitalize on two evolving conditions as they developed:

1. Computers and information technologies (IT) are experiencing substantial advancements in speed, reliability, and power. Individuals in the majority of organizations have grown used, and in many cases, rather intimate, with the benefits of integrated computer-based management systems, including their enhanced speed, accuracy, and capabilities.

2. The process of combining knowledge and decision making in all areas that affect the flow of materials and materials management. This integration encompassed not just the internal functions of marketing, engineering, human resources, accounting, and finance, but also the upstream activities related to supplier information and the downstream activities involved in distribution and delivery. The concept currently acknowledged as supply chain management refers to the process of integrating various movements inside a system.

As the organization's requirements expanded to include a comprehensive approach to materials management, the development of IT systems aligned with those requirements. As these systems expanded in size and complexity compared to the older MRP and MRP II systems, they were renamed as enterprise resource planning (ERP) systems.

ERP is analogous to the MRP II system, with the distinction that it does not focus primarily on manufacturing operations. The entire undertaking is considered. The 14th Edition of the APICS Dictionary describes ERP as a system that establishes a structure for arranging, specifying, and regulating the business procedures required to efficiently manage and direct an organization. This enables the organization to utilize its internal knowledge to gain an advantage in the external environment. In order to function effectively, it is necessary to have applications that cover all levels of a company, including work centers, locations, divisions, and the corporate level. These applications should include features for planning, scheduling, costing, and other relevant functions. ERP is a comprehensive system that includes the entire organization, while MRP II is limited to the manufacturing process. ERP systems have a broad range of functionality that enables the monitoring of orders and other critical planning and control data across the whole organization, spanning from procurement to final customer delivery. Furthermore, numerous ERP systems have the capability to facilitate data sharing across managers across different companies, enabling them to possibly have comprehensive visibility throughout the whole supply chain.

While the strength and capability of these fully integrated ERP systems are indeed significant, they also come with substantial expenditures. Acquiring several of the top-notch systems might be costly. The substantial data demands, encompassing both volume and precision, can result in costly, time-consuming, and challenging implementation of the systems for many companies.

With the expansion of the supply chain idea, a new planning approach was devised. Referred to as **advanced planning and scheduling (APS) systems,** this technique frequently involves incorporating suppliers and consumers into the planning process in order to enhance the overall performance of the supply chain. The system aims to generate a prompt and practical schedule for meeting consumer demand by gathering information from the whole supply chain. These identical principles can also be applied internally within a company's operations to enhance or provide a more practical solution for its customers.

3.13 Making the Production Plan

So far, we have explored the goal, planning horizon, and level of information that are included in a production plan. This section provides specific information regarding the intricacies of formulating production strategies.

The production plan establishes the boundaries or quantities of manufacturing operations for a future period, taking into account the market strategy and the resources already available. It combines the factory's competencies and resources with the market and financial strategies to accomplish the company's entire business objectives.

The production plan establishes the overall levels of production and inventories during the planned period. The primary objective of the production rates is to achieve the goals outlined in the strategic plan and the strategic business plan. These factors encompass inventory levels, backlogs, market demand, customer service, cost-effective factory management, labor relations, and other related aspects. The plan must have a sufficient time horizon to account for the necessary labor, equipment, facilities, and materials required to successfully execute it. Usually, this time frame spans from 6 to 18 months and is divided into monthly intervals, perhaps even weekly ones.

At this level, the planning process disregards specific specifics such as individual items, colors, styles, or options. Due to the lengthy time frames and the unpredictability of demand over extended periods, the level of detail would lack accuracy and practicality, making the plan costly to develop. In order to facilitate planning, it is necessary to have a limited number of product groupings or a single unit.

3.14 Establishing Product Groups

Capacity refers to the inherent capability or potential of an entity to generate or provide products and services. It refers to the presence of adequate resources to meet the level of demand. The duration of a production plan can be defined as either the available time or, occasionally, the quantity of units or dollars that can be produced within a specific timeframe. The demand for commodities must be converted into the demand for capacity. At the production planning level, minimal specificity is necessary. This entails the identification of product groupings, or families, of individual items based on the commonality of their manufacturing processes. For instance, many calculator models may utilize identical procedures and require equivalent capacity, irrespective of any differences between the models. They would be classified as a product family.

Significant alterations to capacity are typically impractical within the timeframe of the production schedule. Modifications or reductions in plant and equipment are either unattainable or exceedingly challenging to execute at this timeframe. Nevertheless, certain aspects can be modified, and it is the duty of industrial management to recognize and evaluate them. Typically, the following can be altered or changed:

- Individuals can be recruited and terminated, additional hours or reduced hours can be worked, and schedules can be modified by adding or eliminating shifts.

- During periods of low demand, inventory can be accumulated and then sold or utilized during periods of strong demand.
- Work can be outsourced or additional equipment might be rented.

Every option has its own advantages and disadvantages. The responsibility of manufacturing management is to identify the most cost-effective option that aligns with the business's aims and objectives.

3.14.1 Basic Strategies

In summary, the production planning problem exhibits the following characteristics:

- A time frame of at least 12 months is utilized, with regular updates occurring on a monthly or quarterly basis.
- Production demand is characterized by a limited number of product families or standardized units. The demand exhibits fluctuations or follows a seasonal pattern.
- Plant and equipment are immovable within the specified period of time.
- Various management objectives are established, including minimizing inventories, optimizing factory operations, providing excellent customer service, and fostering positive worker relations

Figure 3.18 shows a hypothetical demand forecast for a product group. **Note that the demand is seasonal.**

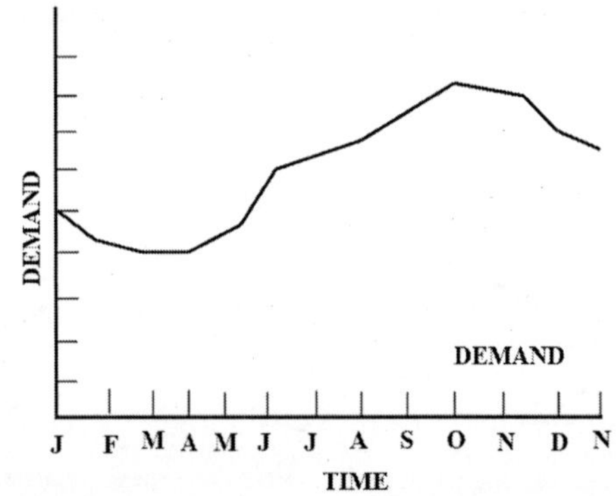

Figure 3.18 Hypothetical Demand Curve

There exist four fundamental tactics that can be employed in the development of a production plan:

1. Chase strategy.
2. Production leveling.
3. Subcontracting.
4. Hybrid strategy.

3.14.1.1 Chase (demand matching) strategy

The chase strategy refers to the practice of manufacturing the exact quantities required at a particular moment. The inventory levels remain consistent while output fluctuates to align with demand. Figure

3.19 depicts this approach. The company produces a sufficient quantity of goods at any given moment to precisely match the demand. For certain sectors, this is the sole approach that can be pursued. Agricultural producers, for example, are required to generate crops within the period of optimal growth. The post office is required to handle mail during both the busy Christmas period and the slower seasons. Restaurants are obligated to provide meals at the desired time of the clients. These industries are unable to accumulate or store their products or services and must have the ability to meet demand in real-time.

Under such circumstances, it is imperative for the company to possess the capacity to effectively fulfill the highest level of demand. During the growing season, farmers require an ample amount of machinery and equipment for harvesting. However, this equipment remains unused during the winter. Businesses that experience fluctuations in demand due to seasonal or cyclical factors frequently need to recruit and provide training for employees during the periods of highest demand, and subsequently terminate their employment once the peak period has ended. Occasionally, it becomes necessary for them to incorporate more shifts and overtime. These modifications incur additional expenses.

The benefit of employing the chase method is the ability to maintain minimal inventory levels. Goods are produced in response to demand and are not stored in large quantities. Consequently, the expenses linked to maintaining inventory are eliminated.

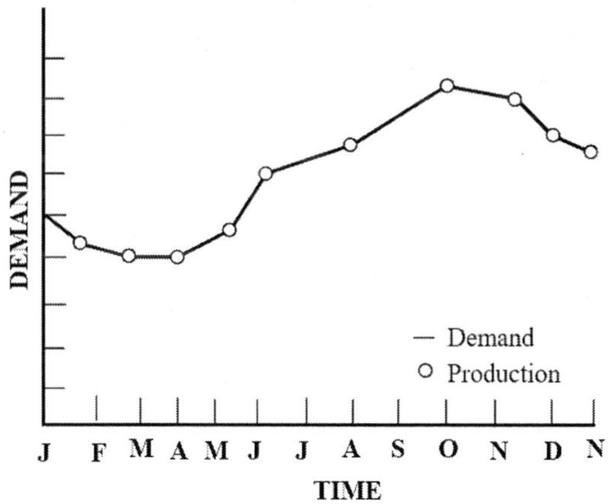

Figure 3.19: Chase (demand matching) Strategy

3.14.1.2 Production leveling

Production leveling involves consistently manufacturing a quantity that matches the average demand. Figure 3.20 illustrates this relationship. Companies determine their aggregate demand over the duration of the plan and, on average, manufacture a sufficient quantity to fulfill it. Occasionally, there is a situation where the quantity of goods produced exceeds the demand, resulting in the accumulation of inventory. During certain periods, there is a higher demand for a product or service, resulting in the depletion of inventory.

A production leveling method has the benefit of maintaining a consistent level of operation, hence avoiding the expenses associated with altering production levels. Companies do not require additional production capacity in order to fulfill high levels of demand. There is no requirement for them to employ and educate personnel and terminate their employment during periods of low demand. They have the ability to establish a reliable and consistent group of employees. An inherent drawback is that during periods of low demand, there will be an accumulation of inventory, resulting in additional expenses for storage and maintenance. Moreover, an underestimated prediction may lead to insufficient production of inventory for the high-demand season. Production leveling refers to the practice of utilizing a company's resources consistently and maintaining a consistent level of output on a daily basis. The precise monthly (or perhaps weekly) production quantity, however, will not remain consistent due to the fluctuation in the number of working days from month to month.

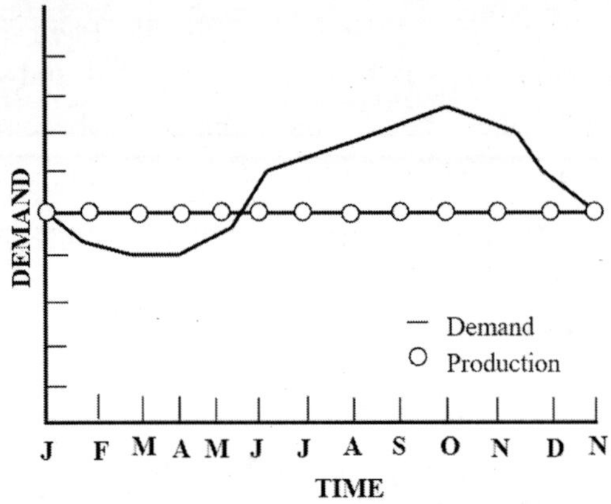

Figure 3.20 Production Leveling Strategy

Example: A company intends to manufacture 10,000 units of an item at a consistent rate over the next three months. The first month comprises 20 working days, the second 21 working days, and the third 12 working days due to an annual closure in the third month. To maintain production at a consistent level, what is the company's average daily production?

Answer: Total production = 10,000 units

Total working days = 20+ 21+ 12 = 53 days

Average daily production $= \frac{10{,}000 \text{ Units}}{53} = 188.7$ units

For highly seasonal products like Diwali lights, implementing a production leveling approach is essential. If a corporation were to implement a chase strategy, it would incur significant expenditures due to idle capacity, as well as expenses related to hiring, training, and laying off employees.

3.14.1.3 Subcontracting

Production leveling, as a pure strategy, entails consistently producing at the predetermined level. Any demand above that level, provided inventory is unavailable, would result in unmet demand or the need to fulfill the additional need through alternative means. One popular manufacturing strategy in such situations is to fulfill any extra demand by subcontracting. Subcontracting refers to the practice of procuring additional quantities from external sources. Another option is to deliberately reject additional demand. The latter can be achieved by price increases during periods of high demand or by lengthening the time it takes to fulfill orders. Figure 3.21 illustrates this scenario.

The primary benefit of these approaches is the reduction in production expenses. By maintaining optimal production capacity, the expenses related to surplus capacity are eliminated. Additionally, the expenditures linked to altering production levels are eliminated due to the consistent production rate. One major drawback of subcontracting is that the expenses associated with purchasing (including item cost, purchasing, transportation, and inspection charges) may exceed the cost of producing the item in-house. Subcontracting can be employed as a component of a chasing strategy.

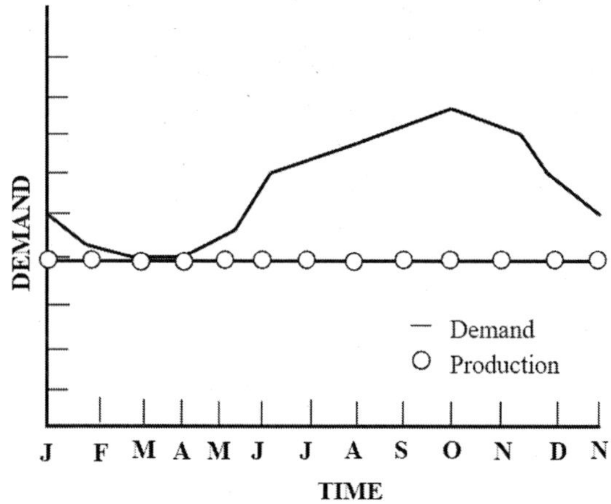

Figure 3.21 Subcontracting

Only a small number of organizations engage in full vertical integration, producing all of their own goods or purchasing all of their required inputs. The determination of whether to purchase or produce certain things mostly hinges on cost, although there are other additional criteria that may be taken into account. Companies may engage in manufacturing in order to safeguard proprietary methods, uphold quality standards, and retain a skilled workforce. They might choose to purchase from a supplier with specialized knowledge in designing and manufacturing a specific component, enabling the company to focus on its own area of competence or obtain reasonable costs. It's easy to make the choice for many things, like nuts and bolts or parts that the company doesn't normally make. For other things that the company is good at, they need to decide if they want to hire or not.

3.14.1.4 Hybrid strategy

The three strategies mentioned thus far are considered pure strategies. Each cost category includes specific expenses: equipment, personnel recruitment/termination, additional work hours, stock, and outsourcing. There are numerous potential hybrid or integrated tactics that a corporation can employ. Each entity will possess its unique set of cost attributes. The role of production management is to identify the optimal combination of techniques that reduces the total costs, meets the needed level of service, and aligns with the objectives of the financial and marketing plans. Figure 3.22 depicts a potential hybrid strategy. There is a partial alignment between demand and production, with some efforts made to even out production levels. Additionally, subcontracting is utilized during the busiest periods. The plan is just one among numerous alternatives that may be established.

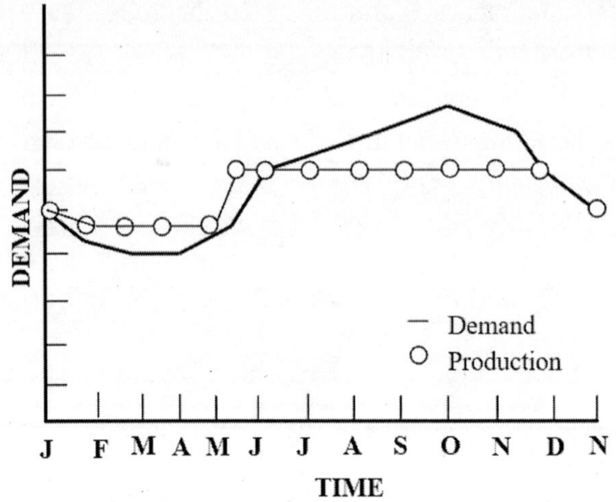

Figure 3.22 Hybrid strategy

3.15 Developing a Make-to-Stock Production Plan

In a make-to-stock setting, items are manufactured and stored in inventory prior to receiving a customer's order. The goods are sold and delivered directly from the inventory. Mass-produced apparel, pre-packaged frozen meals, and bicycles are examples of this type of manufacturing. Typically, companies engage in make-to-stock production when

- Demand is relatively consistent and predictable.
- The selection of products is limited.
- The marketplace demands delivery times that are significantly shorter than the time required to produce the product.
- The product has an extended shelf life.

The information needed to make a production plan is as follows:

- Forecast by period for the planning horizon.
- Opening inventory.
- Desired ending inventory.

Any customer orders that are past due. These are orders that are delayed in delivery and are occasionally referred to as "backorders." As previously mentioned in the chapter, it is important to distinguish between the term's backorders and backlogs. The backlog is a collection of customer orders that have been received but have not yet been shipped, whereas the backorders are orders that are either past due or due for imminent shipment but cannot be fulfilled due to insufficient inventory levels. The goal of creating a production plan is to reduce the expenses associated with inventory management, changing production levels, and stocking out (i.e., not having the products the customer desires at the specified time). The following sections develop a plan for leveling production and one for chase strategy.

3.15.1 Level production plan

Following is the general procedure for developing a plan for level production.

- Aggregate the projected demand throughout the entire planned period.
- Calculate the initial inventory and the target final inventory.
- Determine the total production needed using the following method:
- The formula for calculating total production is as follows: total production equals the sum of the total projection, back orders, ending inventory, minus the starting inventory.
- Determine the necessary production for each period by dividing the total production by the number of periods.
- Determine the final stock of goods for each time period.

Example 3.5: Amalgamated Fish Sinkers specializes in manufacturing a range of high-quality sinkers for fresh fish and is seeking to create a production strategy for these products. The projected initial stock is 100 cases, and the corporation aims to decrease it to 80 cases by the conclusion of the planning term. The number of working days remains constant for each cycle. There are currently no outstanding orders. The projected demand for the fish sinkers is as follows:

Period	1	2	3	4	5	Total
Forecast (cases)	110	120	130	120	120	600

a) What is the optimal production quantity for each period?
b) What is the final inventory for each period?
c) Given that the cost of carrying inventory is Rs. 5 per case every period, what is the overall cost of carrying inventory?
d) What will be the overall expenditure of the plan?

Period		1	2	3	4	5	Total
Forecast (cases)		110	120	130	120	120	600
Production		116	116	116	116	116	580
Ending Inventory	100	106	102	88	84	80	

Table 3.2 Level Production Plan: Make-to-stock

Answer

a) Total production required = 600+ 80-100 = 580 cases

Production each period = 580/5 = 116 cases

b) Ending inventory = opening inventory + production - demand

Ending inventory after the first period = 100+ 116110 = 106 cases

Similarly, the ending inventories for each period are calculated as shown in Table3.2. The ending inventory for period 1 becomes the opening inventory for period 2:

Ending inventory (period 2) = 106 + 116 120 102 cases

c) The total cost of carrying inventory would be = (106102 88 +84 +80) Rs. 5 = Rs. 2300)

d) Since there were no stockouts and no changes in the level of production, this would be the total cost of the plan.

Chase strategy Amalgamated Fish Sinkers also produces a separate range of products known as fish stinkers. Regrettably, these items have a limited shelf life and the company is unable to stockpile them for future sales. They should employ a pursuit strategy and produce only the necessary amount to meet demand in each time period. The inventory costs will be kept to a minimum, and there should be no costs associated with stockouts. Nevertheless, altering output levels will incur expenses.

Assume that in the previous example, increasing or decreasing the output level by one case incurs a cost of Rs. 20. For instance, a transition from 50 to 60 would incur expenses.

(6050) × Rs. 20 = Rs. 200

The initial inventory consists of 100 cases, and the company aims to reduce it to 80 cases throughout the first term. The production necessary for the initial phase would subsequently be required.

110 - (100 - 80) = 90 cases

Given that the output in the period before to period 1 was 100 cases, Table 3.3 illustrates the fluctuations in production levels and ending inventories.

The cost of the plan would be

Cost of changing production level = (60) Rs. 20 = Rs. 1200

Cost of carrying inventory (80 cases) (5 periods) Rs. 5 = Rs. 2000

Total cost of the plan = Rs. 1200 + Rs. 2000 = Rs. 3200

It is important to acknowledge that the previous examples offer a fundamental comprehension of the monetary expense associated with potential production strategies. Prior to making a definitive choice of a plan, it is necessary to assess other factors, some of which are challenging to estimate in terms of financial implications. Some potential concerns may encompass

- The impact on customers due to cyclical demand in a level schedule resulting in shortages.
- The effect on production workers as they are shifted between different production tasks in a chase strategy. This impact frequently diminishes the efficiency of the process.
- Possible reduction in revenue if clients shift their purchasing habits and opt for a competitor's offerings.

Period	0	1	2	3	4	5	Total
Demand (cases)		110	120	130	120	120	600
Production	100	90	120	130	120	120	580
Change in Production		10	30	10	10	0	60
Ending Inventory	100	80	80	80	80	80	

Table 3.3 Chase strategy: Make-to-stock

3.16 Developing a Make-to-Order Production Plan

When a business is make-to-order, they wait for a customer to place an order before they start making the goods. This type of manufacturing includes making clothes that fit perfectly, machines, and anything else that is made exactly how the customer wants it. Things that cost a lot are often made to order. Most of the time, companies make to order when

- Products are manufactured according to customer specifications.
- The customer is patient and willing to endure a wait while the order is being prepared.
- The production and storage costs of the commodity are high.
- Multiple product choices are available.

Assemble-to-order In cases where multiple product choices are available, such as in the automobile industry, and when customers are unwilling to wait for the product to be manufactured, manufacturers build and maintain a supply of standardized component components. Upon receiving an order from a customer, manufacturers gather the necessary component parts from their inventory and assemble them according to the order. Given that the components are readily available, the company merely requires sufficient time to assemble the product before sending it to the consumer. Automobiles and computers are examples of things that are assembled to order. Assemble-to-order is a smaller category within the make-to-order process.

To create a production plan for make-to-order products, the following information is required:

- Periodic forecast for the specified planned horizon.
- Commencing the processing of customer orders that have accumulated over time.
- Targeted final backlog.

Backlog in a make-to-order environment, a company does not accumulate an inventory of completed products. Rather, it is burdened by an accumulation of unfulfilled customer orders. Normally, the backlog pertains to future deliveries and does not encompass orders that are past due or delayed. A custom woodwork business may be occupied for several weeks due to customer orders. This will serve as its backlog. The order will be added to the queue or backlog if individuals desire to have work completed. Manufacturers prefer to manage the congestion in order to deliver exceptional customer service.

Level production plan Following is a general procedure for developing a make-to- order level production plan:

- Aggregate the projected demand throughout the entire planned period.
- Calculate the initial backlog and the target final backlog.
- Determine the total production needed using the following method:

Total production = total forecast + opening backlog - ending backlog

Period		1	2	3	4	5	Total
Forecast (cases)		100	100	100	100	100	500
Planned Production		104	104	104	104	104	520
Projected Backlog	100	96	92	88	84	80	

Table 3.4 Level production plan: Make-to-order

- Determine the necessary production per period by dividing the total production by the number of periods.
- Distribute the current accumulation of unfinished work evenly across the designated time frame based on the deadline for each period.

Example 3.6: A local printing shop offers a personalized printing service. Due to the unique nature of each profession, demand is projected in terms of hours per week. During the upcoming five weeks, the organization anticipates a weekly need of 100 hours. The organization currently has a backlog of 100 hours, and their goal is to decrease it to 80 hours within a span of five weeks. What is the weekly workload required to decrease the backlog? What will be the accumulated amount of unfinished work at the conclusion of each week?

Answer

- Total production = 500 + 100 - 80 = 520 hours
- Weekly production = 520/5 = 104 hours
- The backlog for each week can be calculated as
- Projected backlog = old backlog + forecast – production
- For week 1: Projected backlog = 100 + 100 104 = 96 hours
- For week 2: Projected backlog = 96 + 100 – 104 = 92 hours

Resource Planning

After the basic production plan is created, it needs to be evaluated against the company's current resources. This stage is referred to as resource requirements planning or resource planning. There are two questions that need to be addressed:

1. Do we have the necessary resources to fulfill the production plan?
2. If not, how will the discrepancy be resolved?

If there is insufficient capacity to fulfill the production plan, it is necessary to modify the plan.

An often utilized instrument is the resource invoice. This data represents the amount of essential resources (such as materials, labor, and bottleneck operations) required to produce a single average unit of the product group. Table 3.5 presents a resource bill for a company specializing in the production of tables, chairs, and stools, which are considered as a three-product family.

To determine the required amount of wood and manpower, the company can analyze the production of 500 tables, 300 seats, and 1500 stools within a specific timeframe. For instance, the quantity of timber required is

- Tables: $500 \times 20 = 10{,}000$ board feet
- Chairs: $300 \times 10 = 3000$ board feet
- Stools: $1500 \times 5 = 7500$ board feet
- Total wood required = 20,500 board feet

Product	Wood (board feet)	Labor (standard hours)
Table	20	1.31
Chairs	10	0.85
Stools	5	0.55

Table 3.5 Resource Bill

The amount of labor needed is

- Tables: $500 \times 1.31 = 655$ standard hours
- Chairs: $300 \times 0.85 = 255$ standard hours
- Stools: $1500 \times 0.55 = 825$ standard hours
- Total labor required = 1735 standard hours

Now, the company has to figure out how much wood and work it needs compared to what it can get. Let's say that the average amount of work that can be done during this time is 1600 hours. The top plan needs 1735 hours, which is 135 hours more than the other plan, or 8.4% more. Either more space needs to be found or the main plan needs to be changed. In this case, it might be possible to work extra hours to make up for the lack of capability. If overtime isn't possible, the plan needs to be changed so that less work is needed. This could mean moving some production to an earlier time frame or slowing shipments.

3.17 CAPACITY MANAGEMENT

3.17.1 Introduction

Up until now, our focus has been on planning priority, which involves deciding what should be produced and when. The system follows a hierarchical structure, progressing from long-term planning with limited specifics (production plan), to medium-term timeframes (master production schedule), and finally to a detailed level with short-term timeframes (material requirements plan). At every stage, the production process formulates prioritized plans to meet the demand. Nevertheless, in the absence of the necessary resources to implement the prioritized plan, the strategy will be impracticable. Capacity management focuses on providing the requisite resources. This chapter provides a detailed examination of the concept of capacity, including its definition, the amount that is already accessible, the amount that is needed, and strategies for effectively managing priority and capacity.

3.17.2 Definition of Capacity

Capacity refers to the maximum quantity of work that may be completed during a given timeframe. The 14th edition of the APICS Dictionary defines capacity as the ability of a worker, machine, work center, plant, or organization to provide output within a specific time frame. Capacity refers to the rate at which work is performed, rather than the total amount of work completed. There are two essential types of capacity: available capacity and necessary capacity. Capacity available refers to the ability of a system or resource to generate a specific amount of output within a specified timeframe. Capacity required refers to the amount of system or resource capacity that is necessary to generate a desired output within a specific timeframe. An interconnected concept to capacity required is load. This refers to the total quantity of work that has been assigned to a facility for a specific period of time, including both work that has already been completed and work that is scheduled to be done in the future. The total is the aggregate of all the necessary capabilities. The terms "capacity required," "load," and "capacity available" are crucial in capacity management and will be further examined in upcoming sections of this chapter.

The concept of capacity is commonly represented by a funnel shape, as depicted in Figure 3.23. Capacity availability refers to the speed at which work can be taken out from the funnel. Load refers to the quantity of work present in the funnel.

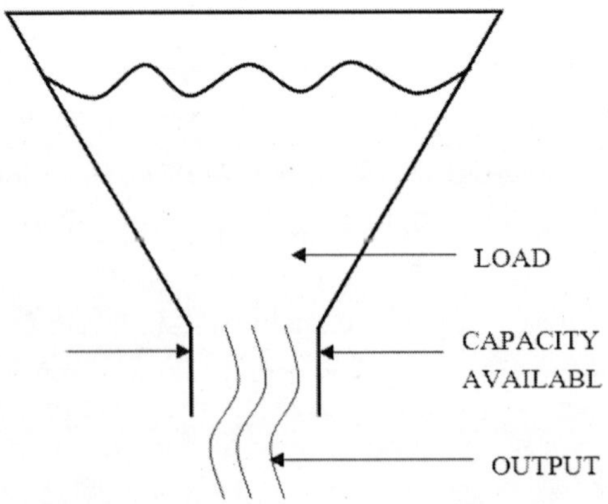

Figure 3.23 Capacity Versus Load

Capacity management is the function that is accountable for assessing the necessary capacity to accomplish the priority plans, and subsequently ensuring, supervising, and regulating that capacity to ensure the priority plan is fulfilled. According to the 14th edition of the APICS Dictionary, capacity management is the process of determining, quantifying, tracking, and modifying the boundaries or levels of capacity in order to carry out all manufacturing schedules. Like any other management procedures, it comprises of the functions of planning and control.

Capacity planning involves identifying the necessary resources and strategies to meet the priority plan and ensure that the requisite capacity is available. It occurs at every stage of the priority planning process. Production planning, master production scheduling, and material requirements planning establish priorities by determining desired items and their respective timelines. However, the implementation of these priority plans is contingent upon the firm having adequate capacity to meet the demand. Capacity planning establishes a connection between the different production priority schedules and the available manufacturing resources.

3.18 Capacity Planning

Capacity planning entails the computation of the necessary capacity to accomplish the priority plan and identifying methods to ensure that capacity is accessible. If the capacity need cannot be fulfilled, the priority plans are unattainable and must be altered.

Priority plans are typically expressed in terms of product units or a standardized unit of output. Capacity can occasionally be expressed using the same units, such as tons of steel or yards of cloth. In the absence of a shared unit, capacity is expressed in terms of the available hours. Subsequently, the priority plan needs to be converted into the number of work hours needed and then compared to the total available work hours. The procedure of capacity planning is as outlined below:

1. Calculate the available capacity at each work center for each time period.
2. Calculate the workload at each work center at each specified time period.
3. Convert the priority plan into the specific number of work hours needed at each work center for each designated time period.
4. To determine the load on each work center in each time period, summarize the capacities needed for each item on each work center.
5. Address disparities between the existing capacity and the necessary capacity. If feasible, modify the existing capacity to align with the load. Alternatively, the priority plans need to be adjusted to align with the current capacity.

This procedure takes place at every stage within the priority planning process, with the only difference being the level of specificity and duration of time involved.

3.18.1 Planning Levels

3.18.1.1 Resource planning

Encompasses the determination of long-term capacity resource needs and is closely connected to production planning. Usually, this process entails converting the product priorities for each month, quarter, or year from the production plan into a comprehensive measure of capacity, such as total gross labor hours. Resource planning include the process of acquiring and eliminating workforce, capital equipment, product design, or other facility adjustments that require a significant amount of time. If it is not possible to formulate a resource strategy that aligns with the production plan, then the production plan must be altered. Both plans establish the boundaries and quantities for production. The master production schedule (MPS) should be attainable if it is based on realistic assumptions.

3.18.1.2 Rough-cut capacity planning (RCCP)

Enhances capacity planning by providing a more detailed analysis. The master production schedule serves as the main source of information for the rough-cut capacity planning (RCCP) process. Rough-cut capacity planning serves the objective of assessing the feasibility of the Master Production Schedule (MPS), issuing alerts for any potential bottlenecks, ensuring optimal usage of work centers, and informing vendors about capacity requirements.

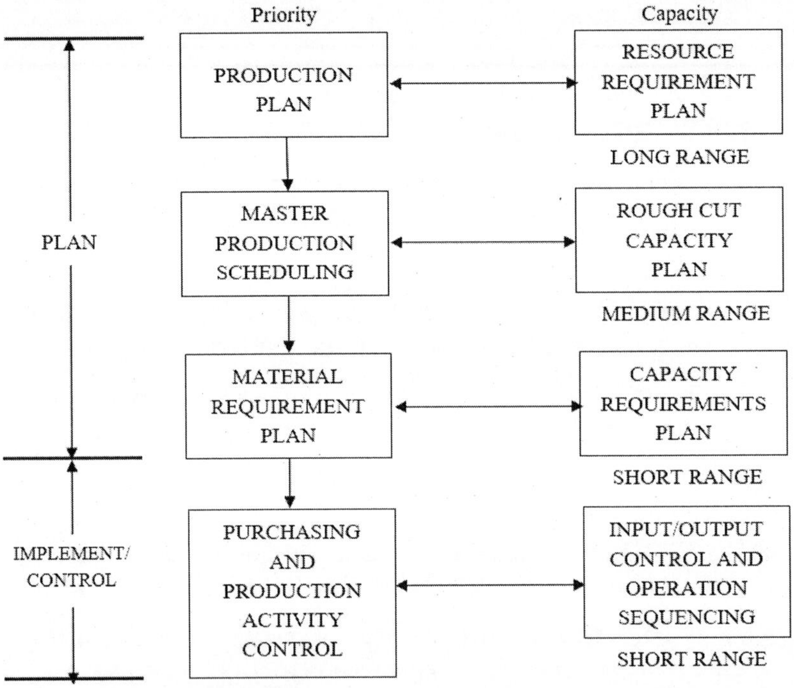

Figure 3.24 Planning Levels

3.18.1.3 Capacity requirements planning (CRP)

Is a process that involves planning for capacity at a more granular level and is closely connected to the material requirements plan. Given its emphasis on individual elements, this form of planning requires a higher level of specificity compared to rough-cut capacity planning. The focus is on individual orders at specific work centers, where it assesses the workload and labor needs for each time period at each work center.

Figure 3.24 illustrates the correlation between various degrees of priority planning and capacity planning. It is important to note that while higher levels of priority planning influence lower levels, the different capacity plans only pertain to their specific level in the priority plan and do not affect following capacity planning levels. Resource planning is a related concept to production planning, although it is not used as an input for rough-cut capacity planning.

Once the plans are finalized, it is necessary to grant authorization for production activity control and purchasing to proceed with processing shop orders and purchase orders. 3.18.2 capacity requirements planning.

The Capacity Requirements Plan (CRP) is executed at the same level as the Material Requirements Plan (MRP). Production analysis involves the meticulous calculation of the specific labor and machine resources required to meet the desired level of production. The MRP generates planned orders and open shop orders, which are then transformed into time-based demand for each work center in every time period. This procedure considers the time it takes for each activity and adjusts the scheduling of operations at work centers accordingly. When assessing open shop orders, it takes into consideration the work that has already been completed on a shop order. Capacity requirements planning is the most meticulous, comprehensive, and precise methodology among capacity planning methods. Precision is crucial, particularly in the near time spans. Due to the level of intricacy, a substantial quantity of data and processing is necessary.

Inputs

Open shop orders, planned order releases, routings, time standards, wait times, and work center capacities are some of the things that CRP needs as inputs. The following facts can be used to get this information:

- Open orders.
- Material requirements plan.
- Routings.
- Work centers.

Open orders: On the material needs plan, an open shop order shows up as a scheduled receipt. It's an approved order for a certain number of parts to be made and delivered on a certain date. It shows all the important details, like amounts, due dates, and activities. All of the open shop sales are kept track of either by hand or in a computer file or table.

Planned order drops with the gross needs for a part as a guide, the computer's MRP logic figures out planned orders. They are used by the CRP process to figure out how much power will be needed in the future.

Routings Work moves from one work area to the next as it is finished. This is called routing. A routing is written on a route sheet, a computer routing file, or a computer routing table. For every part that is made, there should be routing data that includes the following information:

- Tasks that need to be done.
- The order of operations.

- There will be work centers.
- Possible other places to work.
- The tools that are needed for each job.

Standard times include the time it takes to set up and run each piece.

Work centers A work center consists of multiple equipment or workers with the ability to do identical tasks. The equipment will often be comparable, so there are no disparities in the type of tasks the machines can perform or their individual capacities. A collection of sewing machines with comparable capabilities could be regarded as a work center. Work center data encompasses details regarding the center's capacity as well as the durations of its move, wait, and queue periods.

The move time refers to the typical duration required to transfer materials between different workstations. The wait time refers to the duration during which a job remains at a work center following its completion and prior to being relocated. The queue time refers to the duration during which a job is idle at a work center prior to being processed. Lead time is the cumulative duration of queue, setup, run, wait, and move times.

3.19 Capacity Available

The term "capacity available" refers to the ability of a system or resource to generate a specific amount of output within a specified timeframe, as we have previously mentioned. It is influenced by the following factors:

- **Product specifications.** Any alteration in the product requirements might lead to a modification in the manufacturing process and the time needed to produce the product. Consequently, this can impact the quantity of units that can be manufactured within a specified timeframe.
- **Product mix.** Every product has a unique combination of procedures and time constraints involved in its production. Modifying the assortment of products being manufactured can result in a variation in the overall time or capacity needed for the production mix.
- **Plant and equipment.**This pertains to the techniques employed in the production of the product. Modifying the approach, such as utilizing a more efficient machine, can potentially alter the pace of output. Moreover, the addition of extra equipment to the work center will result in a modification of the available capacity.
- **Work effort.** This has to do with how fast or how often the work is done. This is because the capacity will change if the workers speed up and maybe make more in the same amount of time.

Each of these elements has a direct influence on capacity. When there is significant variation, it becomes challenging to utilize product units as a means of measuring capacity. What units are appropriate for measuring capacity?

3.20 Measuring Capacity

Output quantities When the range of items manufactured at a work center or plant is limited, it is often feasible to utilize a standardized unit that applies to all products. Paper mills quantify their capacity in terms of the weight of paper produced, breweries use barrels as a unit of measurement for their production volume of beer, and automobile manufacturers gauge their capacity based on the quantity of vehicles manufactured. However, if a diverse range of items is produced, it may not be possible to find a suitable standard unit of measurement. Time is the universal unit for all products in this scenario.

The term "**standard time**" refers to the amount of time it takes to produce a product using a specific manufacturing procedure. By employing time study methodologies, it is possible to ascertain the standard time, or standard hours, required for a task. This refers to the amount of time it would take an average skilled operator, working at a regular speed, to complete the operation. It serves as a benchmark for quantifying labor and as a standard for expressing volume. Additionally, it is employed in the process of loading and scheduling tasks to be completed.

3.21 Levels of Capacity

Capacity is usually measured on at least three levels:

- Choice between a machine or an individual worker.
- A work center refers to a specific area where work is performed.
- A production line is a sequence of workstations where products are assembled or processed.
- A cell is a self-contained unit inside a production line that focuses on a certain task or process.
- A plant can be defined as a collection of various work centers.

3.21.1 Determining Capacity Available

The three forms of capacity accessible, namely theoretical, calculated or rated, and demonstrated or measurable, are determined based on the criteria of available time, usage, and efficiency.

Time availability: - The available time refers to the total number of hours that a work center can be utilized. For instance, a labor center that operates for one 8-hour shift, five days a week, is accessible for a total of 40 hours each week. The amount of time that is accessible is contingent upon the quantity of machines, the quantity of personnel, and the duration of operation.

Example 3.7: A work center is operational for eight hours per day, five days per week, and comprises three devices. What is the available time?

Answer: Available time = 3 x 8 x 5= 120 hours per week

Utilization This is the utmost number of hours that the work center can provide. On the other hand, it is improbable that this will be achieved consistently. Machine breakdowns, absenteeism, training, material shortages, and other causes of unavoidable delays or inactive time can all contribute to downtime. It is necessary to account for the time spent on meals or breaks if the machine's operation is also contingent upon an individual. The utilization of a work center is the ratio of the active time to the available time: the number of hours that are actually performed. Applications. Utilization.

$$\text{Utilization} = \frac{\text{hours actually worked}}{\text{available hours}} \times 100\%$$

Example 3.8: A work center is available 120 hours but actually produces goods for only 100 hours. What is the utilization of the work center?

Answer: Utilization = $\frac{100}{120} \times 100\% = 83.3\%$

Utilization can be determined from historical records or by a work sampling study.

3.22 Efficiency

A work center has the capacity to operate for 100 hours per week, although it may not generate 100 standard hours of work. Efficiency quantifies the ratio of production to a predetermined benchmark. The workers may be operating at a higher or lower rate than the typical working pace, resulting in the work center's efficiency being either above or below 100%. For instance, during a specific work period, the anticipated result could be 100 hours or the manufacturing of 50 pieces. However, the final result is 120 hours, resulting in the production of 60 units.

$$\text{Efficiency} = \frac{\text{standard hours produced}}{\text{hours actually worked}} \times 100\%$$

Example 3.9: A work center produces 120 standard hours of work in 100 hours. What is the efficiency?

Answer: Efficiency = $\frac{120}{100} \times 100\% = 120\%$

3.23 Productivity

APICS Dictionary, 14th edition defines productivity as "the overall measure of the ability to produce a good or a service." It compares the actual output of production to the actual input of all resources, incorporating the utilization of the time available and the efficiency during that time. There are many ways of measuring productivity. An example of a productivity calculation that can be applied to each work center uses the following formula:

Productivity = utilization X efficiency × 100%.

Example 3.10: A work center operates 32 hours of the 40 hours available. During that time, they produce output equivalent to 30 hours of work.

Answer: Productivity = $\frac{32}{40} \times \frac{30}{32} \times 100\% = 75\%$

Theoretical capacity refers to the greatest capacity that is achievable without considering factors such as downtime, utilization, or efficiency. If a corporation employs two 8-hour shifts at a labor facility, the theoretical daily capacity would amount to 16 hours.

Rated capacity The available capacity at a work center for a period of time, and accounting for the average utilization and efficiency of that work center, is known as calculated or rated capacity.

Rated capacity = available time X utilization X efficiency

Example 3.11: A work center comprises four machines and operates for eight hours per day, five days a week. In the past, the utilization rate has been 85% and the efficiency rate has been 110%. What is the maximum capacity?

Answer: Available time = 4 x 8 x 5 = 160 hours per week

Rated capacity 160 x 0.85 x 1.10 149.6 standard hours

The expectation is to produce 149.6 standard hours of work from that work center in an average week.

3.24 Demonstrated Capacity

A work center comprises four machines and operates for eight hours per day, five days a week. In the past, the utilization rate has been 85% and the efficiency rate has been 110%. What is the maximum capacity?

Example 3.12: During the past 4 weeks, a work center has generated 120, 130, 150, and 140 standard hours of work. What is the proven capability of the work center?

Answer: Demonstrate capacity $= \frac{120+130+150+140}{4}$ 135 standard hours

It should be noted that the shown capacity refers to the average output, not the maximum output. The utilization and efficiency of the work center are important factors, but they are not considered in the calculation because they are already accounted for in the production records utilized for the calculation. Efficiency and utilization can be derived from historical data by keeping track of the hours available, hours worked, and the standard hours produced by a work center.

Example 3.13: During a 4-week duration, a work center completed 540 standard hours of work, had a total availability of 640 hours, and worked for a total of 480 hours. Determine the usage and efficiency of the work center.

Answer:

$$\text{Utilization} = \frac{\text{hours actually workd}}{\text{available hours}} \times 100 \times \frac{480}{640} \times 100\% = 75\%$$

$$\text{Efficiency} = \frac{\text{standard hours of work produced}}{\text{hours actually worked}} \times 100 \times \frac{540}{480} \times 100\% = 112.5\%$$

UNIT SUMMARY

This unit is for engineering diploma students and looks at a number of important business and marketing topics. It starts by looking at the complicated link between price and worth from the customer's point of view. This shows how important it is to understand how customers think when setting prices. From here, students learn about the complicated web of costs that come with running a business and making a new product. This network has many types of groups, such as set, changeable, and overhead costs. By going over the basics of break-even analysis in more detail, the unit also gives students the tools they need to figure out how profitable an idea is and make smart choices. Furthermore, it gives a complete picture of how to create a strong marketing plan, especially when it comes to engineering projects, by making goals and strategies clear. Utilizing real-world examples and clear explanations, students learn how to easily apply these business concepts to their engineering studies. We discussed about the production planning and control (MPC) method in this chapter. It starts by going over the whole system and then gets into the details of making plans for production. After that, you can talk about master orders, planning for material needs, managing capacity, keeping an eye on production, and buying. In order to plan and control output, the first step is to make plans for it. In general, one year is enough time to plan. The shortest possible time frame is based on how long it takes to get the materials and make the product. A lot of information is missing. Plans are often made for groups of things that are made the same way or with the same unit. When you plan for sales and management, you also plan for production. This is a planning process at the executive level where different company units or jobs have to make trade-offs.There are three main ways to make a production plan: go after, level up production, or mix the two. When it comes to prices and how they work, each has pros and cons. The people in charge of manufacturing have to think about how to best combine these basic plans to keep costs low and customer service good. The unit also includes a lot of different questions and tasks that make it easier for students to use what they have learned. Additionally, it offers many chances for evaluation.

EXERCISES NO: -3

Multiple Choice Questions

1. Which of the following best defines the concept of price?
   a. The monetary value of a product or service
   b. The perception of quality associated with a product
   c. The promotional activities undertaken by a company
   d. The distribution channels used to reach customers
2. How does a customer's perception of value influence their purchasing decisions?
   a. It determines the advertising budget of a company
   b. It affects the pricing strategy of a product
   c. It influences the choice of distribution channels
   d. It determines the size of the product
3. What do company and product costs primarily include?
   a. Costs associated with marketing activities
   b. Costs related to research and development
   c. Costs incurred during production and operation
   d. Costs of legal compliance
4. Which of the following is NOT a type of cost typically encountered in business operations?
   a. Fixed costs
   b. Variable costs
   c. Indirect costs
   d. Static costs
5. What is the purpose of a break-even analysis?
   a. To maximize profits
   b. To minimize costs
   c. To determine the point at which revenue equals expenses
   d. To forecast market demand

6. Which tool is commonly used to represent break-even analysis graphically?
   a. Pareto chart
   b. Scatter plot
   c. Gantt chart
   d. Break-even chart

7. Break-even analysis is particularly useful for which of the following decision-making processes?
   a. Selecting distribution channels
   b. Choosing suppliers
   c. Determining production volumes
   d. Setting advertising budgets

8. What does the overall marketing strategy of a company encompass?
   a. Financial forecasting
   b. Customer service policies
   c. Advertising campaigns
   d. Market segmentation and positioning

9. In break-even analysis, what does the term "break-even point" represent?
   a. The point where total revenue equals total variable costs
   b. The point where total revenue equals total costs
   c. The point where total revenue exceeds total costs
   d. The point where total variable costs equal total fixed costs

10. Which of the following statements best describes the purpose of break-even analysis?
   a. To determine the maximum possible profit for a product or service
   b. To calculate the minimum selling price required to cover costs
   c. To identify the level of sales needed to cover all costs
   d. To forecast future market demand for a product or service

11. In break-even analysis, which of the following factors directly affects the break-even point?
   a. Variable costs per unit
   b. Fixed costs
   c. Selling price per unit
   d. All of the above

12. What does the slope of the total cost line indicate on a break-even chart?
   a. The level of fixed costs
   b. The level of variable costs
   c. The profit margin per unit
   d. The rate of change in total costs

13. Which of the following statements is true regarding the break-even point?
   a. It represents the point where total revenue equals total fixed costs.

b. It is independent of changes in variable costs.
c. It indicates the level of sales needed to cover both fixed and variable costs.
d. It is not affected by changes in selling price.

14. How does an increase in fixed costs affect the break-even point?
a. It shifts the break-even point to the right.
b. It shifts the break-even point to the left.
c. It does not affect the break-even point.
d. It depends on the change in variable costs.

15. What does the intersection point of the total revenue and total cost lines represent on a break-even chart?
a. The break-even point
b. The point of maximum profit
c. The point of minimum loss
d. The point of zero sales

16. What is the primary objective of aggregate production planning?
a. Minimize inventory costs
b. Maximize production efficiency
c. Meet customer demand while minimizing costs
d. Reduce lead times

17. When adjusting capacity to meet demand in aggregate production planning, what strategy involves hiring and firing workers to match production levels with demand?
a. Level production strategy
b. Chase demand strategy
c. Subcontracting strategy
d. Overtime strategy

18. Demand management in aggregate planning involves:
a. Forecasting future demand
b. Adjusting production capacity
c. Scheduling production activities
d. Allocating resources effectively

19. What is the key difference between hierarchical and collaborative planning?
a. Hierarchical planning involves top-down decision-making, while collaborative planning involves bottom-up decision-making.
b. Hierarchical planning focuses on short-term goals, while collaborative planning focuses on long-term goals.
c. Hierarchical planning is centralized, while collaborative planning involves input from multiple stakeholders.
d. Hierarchical planning is manual, while collaborative planning uses automated systems.

20. Aggregate planning for services often involves:

a. Adjusting inventory levels
b. Hiring and training staff
c. Procuring raw materials
d. Optimizing production schedules

Answers of Multiple-Choice Questions

1 a, 2 b, 3 c, 4 d, 5 c, 6 d, 7 c, 8 d, 9 b, 10 c, 11 d, 12 b, 13 c, 14 b, 15 a, 16 c, 17 b, 18 a, 19 c, 20 b.

Short and Long Answer Type Questions

1. What does Price mean? And talk about how important price is in today's world of quick changes.
2. Talk about how important it is to know how customers see value when setting prices.Name and explain the two ways of setting prices based on value.
3. Explain the term Break-Even Point theory. How does it help in equipment replacement decisions?
4. A ceiling fan manufacturer buys a component at Rs. 10 each. If he decides to make it himself, his fixed and variable cost will be Rs. 10,000 and Rs. 5 per component respectively. To decide whether to make or buy the component, calculate B.E.P.
5. Explain the term break-even point. Define 'angle of incidence' and 'margin of safety?
6. Write short notes on:
   - Non-linear break-even analysis.
   - Break-even point analysis, a tool for decision-making.
7. The annual sales of a company is Rs. 120,000. Fixed costs including insurance, depreciation etc. is Rs. 15,000. Variable costs Rs. 3 per unit. Plant capacity was 40,000 units and the selling price is Rs. 4 per unit. Total investment in plant and equipment was Rs. 110,000.
   - Draw a break-even for the company.
   - Find and show the profit before taxes, if any, and compute the rate of return on the investment.
8. A concern sells 3000 cycles every year - the price of each is Rs. 600. The available cost data are shown below:

Material cost = Rs. 400,000
Labour cost = Rs. 300,000
Factory cost = Rs. 250,000 (Fixed)
= Rs. 350,000 (Variable)
Administration and distribution overhead
Rs. 150,000 (Fixed)
Rs. 100,00 (Variable)

Draw the break-even chart to scale and mention the minimum economic production. What will be the profit, if the selling price is reduced by 5% without any reduction in the production cost? Also show the new Break-even point and plot the new unit cost curve.

9. The following cost figures are taken from the budget of a company producing consumer goods (budget period is one year)

Direct material	= Rs. 1,800,000
Direct labour	= Rs. 1,600,000
Factor overhead	= Rs. 900,000 (Fixed) + Rs. 400,000 (Variable)
Administrative overhead	= Rs. 200,000 (Variable) + Rs. 600,00 (Fixed)
Selling and distribution overhead	= Rs. 500,000 (Variable) + Rs. 300,00 (Fixed)

In the sales budget the sales amount for the budgeted period is estimated at Rs. 6,100,000.

- Draw the break-even chart.
- What is the estimated profit?
- What is the sales amount at the break-even point?
- How in the chart the result of a 10% increase in the cost in the product selling price without any change

10. You are a production manager at a large automobile manufacturing plant in Gujrat, specializing in producing a variety of vehicle models. With increasing competition and consumer demand for customization, your company is facing challenges in meeting production targets while ensuring high-quality standards and efficiency. In this scenario, discuss the critical need for which strategy in automobile manufacturing operations. Consider factors such as the complexity of assembly tasks, the demand for precision engineering, and the importance of minimizing errors and defects in the final products. Explore different types of your strategy and method of implementation in company.

KNOW MORE

REFERENCES

[1] *Cheser, R. (1994) Kaizen is more than continuous improvement. Quality Progress 27, 23-26.*

[2] *Drucker, P. (1991) The new productivity challenge. Harvard Business Review 69, 69-79.*

[3] *García, Javier, et al. & quot.;Challenges and Opportunities in Part Print Analysis for Aerospace Applications.& quot.; Aerospace Science and Technology, vol. 96, 2021, 106114.*

[4] *Gunasekaran, A., Goyal, S. K., Martikainen, T. and Yli Olli, P. (1993) Modelling and analysis of just- in-time manufacturing systems. International Journal of Production Economics 32, 23-37. Gunasekaran, A., Korukonda, A. R., Virtanen, I. and Yli-Olli, P. (1994) Improving productivity and quality in manufacturing organizations. International Journal of Production Economics 36, 169-183.*

[5] *Ross, K. W. (1994) Assembly-line job satisfaction and productivity. Industrial Engineer 26, 44–45.*

[6] *Rossler, P. E. and Beruvides, M. G. (1994) Management déjà vu? Scientific and total quality management. Engineering Management Journal 6, 6-15.*

[7] *Schoenberger, R.J. (1982) Japanese Manufacturing Techniques: Nine Hidden Lessons in Simplicity. Free Press, New York.*

[8] *Shapiro, B. P. (1977) Can marketing and manufacturing co-exist? Harvard Business Review 55, 104–114.*

[9] *Shiote, Y., Akiyama, J., Yoshida, H. and Harada, Y. (1995) Development of intelligent automated assembly technique. In: Proceedings of the IEEE International Conference on Robotics and Automation, IEEE Press, Piscataway, NJ, Vol. 1, pp. 788-793.*

-----------------*---------*

4 ASSEMBLY LINE BALANCING

UNIT SPECIFICS

The following topics are covered in this unit:

- Assembly lines
- Assembly line balancing
- Splitting tasks
- Flexible and U-shaped line layouts
- Mixed model line balancing
- Current thoughts on assembly lines
- Computerized assembly line balancing.

All the topics are discussed in a lucid manner along with their engineering applications. For student's practice, a moderate number of multiple-choice questions, short and long answer type questions and numerical problems are also given.

RATIONALE

In this unit, a comprehensive exploration of assembly lines is provided, covering various essential aspects crucial to engineering applications. Beginning with the fundamentals of assembly lines, the unit delves into assembly line balancing, elucidating techniques to optimize task allocation and achieve efficient production processes. Concepts such as splitting tasks, flexible and U-shaped line layouts, and mixed model line balancing are thoroughly discussed, offering insights into their practical implementation and benefits. Additionally, the unit delves into current thoughts on assembly lines, incorporating contemporary perspectives and advancements in the field. A diverse range of exercises, including multiple-choice questions, short and long answer type questions, and numerical problems are provided which ensure a comprehensive understanding and mastery of the subject matter

PRE-REQUISITES

- Fundamental Knowledge of any Manufacturing Industry Layout
- A course on Industrial Engineering

UNIT OUTCOMES

After understanding this chapter students will be able to

U4-O1: Understand the perception of Assembly Line in Industry

U4-O2: Understand the Historical concept and types of assembly line used in Industry.

U4-O3: Understand the importance of Line balancing of assembly line.

U4-O4: Understand the Concept Mixed model line balancing.

U4-O5: Understand the different case studies of mix model assembly line.

Unit-4 Outcomes	EXPECTED MAPPING WITH COURSE OUTCOMES (1- Weak Correlation; 2- Medium correlation; 3- Strong Correlation)				
	CO-1	CO-2	CO-3	CO-4	CO-5
U4-O1	3	1	3	3	0
U4-O2	3	1	3	2	0
U4-O3	3	1	2	3	0
U4-O4	3	0	3	2	0
U4-O5	3	0	3	3	0

4.1 Assembly Lines

Most consumer products are assembled during the manufacturing process. Every item consists of multiple components that are joined together using various techniques. These items are often produced using manual assembly lines. The utilization of manual assembly lines is strengthened by the subsequent factors:

- There is either a high or medium level of demand for the product
- Similar products are being produced on the production line.
- Product assembly can be broken down into smaller tasks.
- Automation of assembly processes is not viable due to technological or economic constraints.

Table 4.1 lists products that typically come from manual assembly lines and are characterized by these characteristics.

Some of the reasons why hand assembly lines are more productive than alternative techniques in which several individuals complete all of the duties necessary to construct the items.

- Specialization of work. Adam Smith's "division of labor" principle states that when a huge job is divided into tiny parts and assigned to one worker, the worker becomes extremely capable of executing the single task. Every employee becomes a specialist.
- Interchangeable parts are constructed with close tolerances, allowing for assembly with any part of the same type. Without interchangeable parts, assembly would necessitate filing and fitting of mating components, making assembly line methods impossible.
- The work flow principle involves delegating tasks to workers rather than the other way around. Each work unit runs seamlessly through the manufacturing line, covering the shortest distance between locations.
- Line pacing. Workers on an assembly line are typically required to accomplish their assigned responsibilities on each work unit within a specific cycle time, allowing the line to maintain a consistent output rate. Pacing is typically done using an automated conveyor.

In the current chapter, the engineering and technology of manual assembly lines are conferred.

Audio equipment	Electric motors	Pumps
Automobiles	Furniture	Refrigerators
Cameras	Lamps	Stoves
Cell phones and smart phones	Luggage	Tablet computers
Cooking ranges	Microwave ovens	Telephones
Dishwashers	Personal computers	Toasters and toaster ovens
Dryers (laundry)	Peripherals (keyboards)	Trucks, light and heavy
DVD players	Printers, Monitors, etc.	Video game consoles
E-Book readers	Power tools (drill, saws, etc.)	Washing machine (laundry)

Table 4.1 Items Typically Produced via Manual Assembly Lines

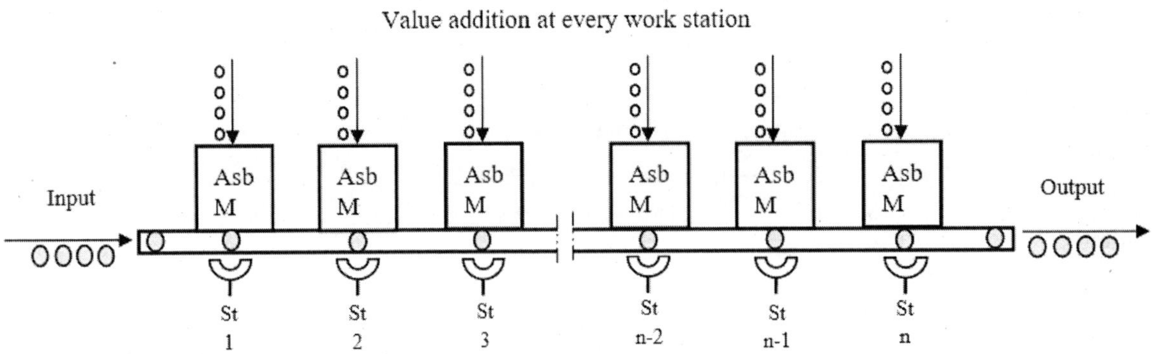

Asb = assembly, M = manual, St= workstation, n = number of stations on the line.

Figure 4.1 Diagram of a Manual Assembly Line

4.2 Introduction of manual assembly lines

A manual assembly line is a production line that, as Figure 4.1 shows, consists of a series of workstations where human workers carry out assembly activities. Assembly of the products takes place as they pass along the line. At every station, a worker completes a percentage of the unit's overall job. Base parts are often "launched" at the start of the line at regular intervals. As each foundational piece passes through a station, workers add components to progressively assemble the finished product. Usually, the base parts are moved along the line by an automated material transport system as they are progressively converted into finished products. The slowest station on an assembly line dictates its output rate. The slowest station eventually limits stations that are capable of functioning faster.

As pointed up in point No. 4.2, manual assembly line technology played a major part in the growth of American manufacturing in the 20th century. It is still a vital manufacturing system used globally to make the assembled goods shown in Table 4.1, as well as cars and consumer appliances.

4.2.1 Backgrounds of the Manual Assembly Line

The foundation of manual assembly lines is mainly two basic labor ideas. Adam Smith (1723–1799) put out the idea, division of labor, in his 1776 English book The Wealth of Nations. The book illustrates the division of labor with a pin factory, showing how 10 workers, each with a specialty in the many different activities needed to make a pin, produced 48,000 pins a day. In contrast, typical production produced only few pins a day and required each worker to do every operation on each pin. Though division of labor had been used in Europe for centuries before Smith, he was the first to recognize its importance in production.

The second work concept, developed at the start of the 19th century by Eli Whitney (1765–1825) and others, is replaceable parts. Assembly line technology would not be conceivable without interchangeable parts.

The meat business in Cincinnati, Ohio and Chicago, Illinois is where contemporary production lines first appeared. The meat packing operations of the mid- and late 1800s moved the butchered animals from one worker to the next using unpowered overhead conveyors. Later, power-driven chain conveyors were installed to construct "disassembly lines," the forerunner of the assembly line. The division of work allowed each meat cutter to focus on a specific task.

American automaker Henry Ford had seen these meat packaging plants. His technical colleagues and he planned an assembly line in Highland Park, Michigan, in 1913 to make magneto flywheels. Efficiency quadrupled. Ford, swept up in success, used assembly line methods to chassis construction. Using chain-driven conveyors and workstations set up to suit the comfort and convenience of his assembly line workers, he was able to double production over earlier single-station assembly techniques. The primary offering of Ford Motor Company at the time, the Model T Ford, saw dramatic price drops as a result of these and other advancements. Ford's success at cost reduction allowed a large number of Americans to now purchase a car. This promoted more research and application of automated transfer lines and other production line methods. It also made suppliers and rivals of Ford copy his techniques, and the manual assembly line became a staple of American business.

4.2.2 Assembly Workstations

On a manual assembly line, a workstation is an area chosen along the work flow path where one or more workers complete one or more job elements. The work aspects stand in for discrete portions of the overall effort required to put the product together. Table 4.2 lists typical assembly tasks carried out at stations in a manual assembly line. A given workstation also has the equipment (powered or hand) needed to complete the task allocated to it.

While some workstations let employees to sit, others are designed for them to stand. Workers that are standing can move about the station area carrying out their designated duties. This is typical for big product assembly, including cars, trucks, and big appliances. Usually a conveyor moves the goods through the station at a steady speed. The worker walks back to the next work unit and restarts the cycle after starting the assembly process close to the upstream side of the station and moving with the work unit until the task is finished. Typically, workstations for smaller completed products—like tiny appliances, electronics, and subassemblies used on larger products—are made so that employees may sit while they work. This is often more suited to accuracy and precision in the assembly job and is more comfortable and less exhausting for the workers.

Application of adhesive	Expansion fitting applications	Snap fitting of two parts
Application of sealant	Insertion of components	Soldering
Arc welding	Press fitting	Spot welding
Brazing	Printed circuit board assembly	Stapling
Cotter pin applications	Riveting and eyelet applications	Stitching
Crimping	Shrink fitting applications	Threaded fastener applications

Table 4.2 Assembly Processes Executed on a Manual Assembly Line

4.2.3 Work Transport Systems

Work units can be moved along a manual assembly line in two fundamental ways: either manually or by a mechanical system. Both approaches offer the fixed routing—that is, all work units move through the same set of stations—that is typical in manufacturing lines.

Manual Methods of Work Transport When it comes to manual work transport, the units of product are moved from one station to another by the workers themselves. Starvation and blockage are the two issues that arise as a consequence of this way of operation. The circumstance known as "starving" occurs when the assembly operator has finished the task that was assigned to the current work unit, but the next unit has not yet arrived at the station. Therefore, the worker is in desperate need of job. The operator has finished the task that was assigned to the present work unit, but they are unable to move the unit on to the downstream station since the worker who is supposed to receive it is not yet prepared to do so. This is what is meant by the term "blocking." Because of this, the operator is unable to do their duties.

It is occasionally necessary to use storage buffers between stations in order to mitigate the harmful consequences of these issues. In certain instances, the work units that are produced at each station are gathered together in batches before being sent to the subsequent station. There are also situations in which work units are carried separately along a flat table or because there is no powered conveyor. All that is required of the worker is to push the unit in the direction of the downstream station once the assignment has been completed at each station. In front of each workstation, there is typically room for one or more work units to come into existence. Not only does this provide the station with a readily available supply of work, but it also makes place for finished units that come from the station further upstream. As a result, starvation and blocking are reduced to a minimum. One of the problematic aspects of this kind of operation is that it has the potential to generate a substantial amount of work-in-process, which is economically undesirable. on addition, workers on lines that rely on manual transport methods are not expected to pace themselves, and the output rates are typically lower.

Transportation for Mechanized Work The movement of units along manual assembly lines is often accomplished through the utilization of powered conveyors and many other types of mechanical material handling equipment. It is possible to construct these systems in such a way that they can enable timed or unpaved functioning of the line. Work transport systems in production lines can be broken down into three primary categories: (a) continuous transport, (b) synchronous transport, and (c) asynchronous transport. The diagrammatic representation of these can be found in Figure 4.2. The following table provides a classification of some of the material transport equipment that is typically associated with each of these categories respectively.

One example of a continuous transport system is shown in Figure 4.2(a), which depicts a conveyor that moves in a continuous motion and maintains a constant velocity. It is typical practice for hand assembly lines to use this strategy. Typically, the conveyor will go the entire length of the line being used. On the other hand, if the line is extremely lengthy, as it would be in the case of a facility that specializes in the final assembly of automobiles, it is segmented into sections, and each section has its own individual conveyor.

There are two methods that can be utilized to carry out continuous transport: (1) work units are attached to the conveyor, and (2) work units can be removed from the conveyor, respectively. Because of its size and weight, the product in the first scenario is unable to be taken from the conveyor. Examples of such products include automobiles and washing machines. In order for the workers to successfully complete the work that has been assigned to them, they will need to walk alongside the product at the same speed as the conveyor.

In the event that the work units are of a small size and light weight, it is possible to remove them from the conveyor in order to make the operator's physical experience at each station more comfortable. An additional convenience for the worker is that the task that has been allocated to them at the station does not have to be finished within a predetermined amount of time. When it comes to dealing with any technological issues that may arise with a certain work unit, every worker is equipped with the freedom to do so. On average, however, it is necessary for each worker to keep up a production rate that is comparable to that of the other workers in the line. In the event that this does not occur, the line will generate incomplete units. Complete Assembled units are produced when parts that were intended to be added at a station are not added because the worker at that station ran out of time.

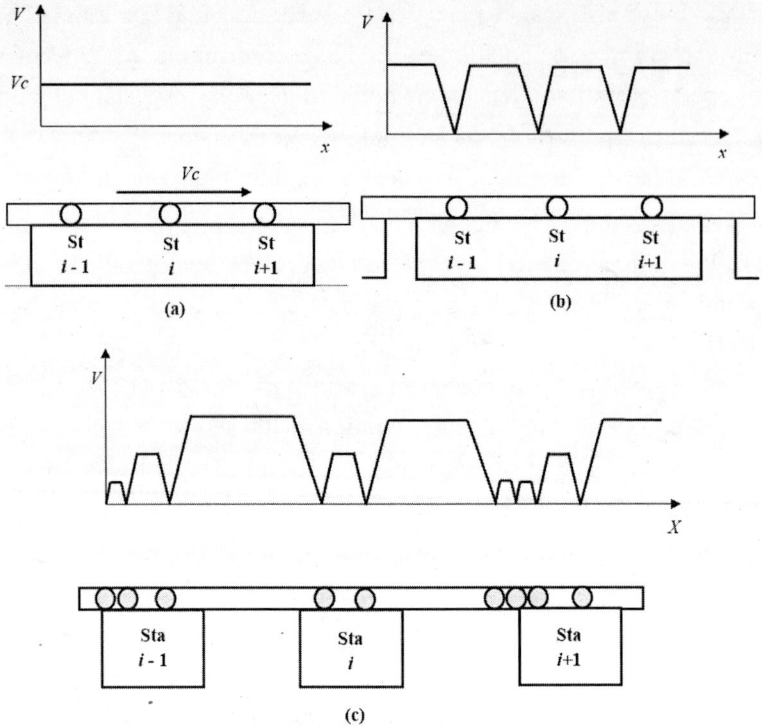

Figure 4.2 Velocity-distance diagram and physical layout for three types of mechanized transport systems used in production lines: (a) continuous transport, (b) synchronous transport, and (c) asynchronous transport.

v = velocity, ve = constant velocity of continuous transport conveyor, x = distance in conveyor v = velocity, ve = constant velocity of continuous transport conveyor, x = distance in conveyor direction, Sta workstation, i workstation identifier.

Work Transport System	Material Handling Equipment
Continuous transport	Trolley conveyor
	Belt conveyor
	Roller conveyor
Chain conveyor	
Synchronous transport	Walking beam transport equipment
	Rotary indexing mechanisms
Asynchronous transport	Overhead conveyor
Track conveyor	
	Automated guided vehicle
	Monorail systems
	Chain/belt driven systems

In **synchronous transport systems**, all work units are transported simultaneously between stations with a rapid, unpredictable motion, and then placed at their corresponding stations. Figure 4.2(b) illustrates a system referred to as intermittent transport, which describes the movement encountered by the work units. Manual lines do not commonly use synchronous transport since it necessitates completing the work within a specific time constraint. This might lead to undue strain on the assembly workers and lead to the production of unfinished products. Although synchronous transport may have drawbacks for manual assembly lines, it is frequently well-suited for automated production lines, where mechanized workstations function at a consistent cycle time.

In an **asynchronous transport system**, a work unit departs from a specific station once the specified duty has been completed and the worker releases the unit. Work units operate autonomously, rather than in a synchronized manner. At any given time, certain units are in transit between workstations while others are currently positioned at stations, as depicted in Figure 4.2(c). Asynchronous transit systems allow for the formation of tiny lines of work units in front of each station. This approach is generally lenient towards fluctuations in the duration of worker tasks.

4.2.4 Line Pacing

A manual assembly line operates at a specific cycle time that is acknowledged to meet the necessary production rate of the line. The computation of this duration is elucidated in Section 4.3. Each worker is expected to perform their allocated task within the cycle time in order to meet the required production rate. The success of a manual assembly line is attributed to the efficient pace at which the people operate. Pacing instills a sense of order and control among assembly line workers, ensuring a consistent and predetermined rate of production. From a managerial perspective, this is advantageous.

Manual assembly lines can be organized with three different levels of pace:

- **Rigid pacing**
- **Pacing with margin**
- **No pacing.**

Under rigid pacing, each worker is allocated a specific and unchanging amount of time during each cycle to finish the specified task. The permitted duration is carried out by a synchronized work transportation system and is typically equivalent to the cycle duration of the line. The presence of rigid pacing has two unfavorable characteristics, as previously stated. Initially, adhering to a strict pace places significant physical and emotional strain on human workers. While a certain amount of stress might enhance human performance, the rapid pace of working on an assembly line for an extended 8-hour shift or more can have detrimental consequences on workers. Furthermore, in a strictly regulated process, if the task is not finished within the predetermined time frame, the work unit leaves the station without being fully completed. This could impede the progress of subsequent jobs at stations farther down the line. Any unfinished tasks on the work unit at the regular workstations must be completed by another worker in order to produce a satisfactory product.

In pacing with margin, the worker is given a specific time frame to accomplish the task at the station, in order to maintain a margin of pace. The maximum duration of the range exceeds the cycle time, allowing a worker to take additional time in case of any issues or if the task time for a specific work unit is higher than the average. This situation arises when multiple product styles are being manufactured on the same assembly line. There are multiple methods to achieve pace with margin: (1) Facilitating the formation of work unit queues between stations, (2) constructing the line in a way that ensures the time a work unit spends within each station exceeds the cycle time, and (3) enabling the worker to move outside the confines of their designated station. Method (1) utilizes an asynchronous transit system to implement work units. These work units are permitted to create queues in front of each station, ensuring that the workers are always provided with work and preventing them from being deprived of tasks. Additionally, this setup allows for some work units to have additional time if other units are completed more quickly. Method (2) is applicable to situations when work units are affixed to a continuously moving conveyor and cannot be detached. Due to the constant conveyor speed, if the station length exceeds the distance required for the worker to finish the assigned task, the work unit will spend more time within the station borders (known as tolerance time) than the cycle time. In method (3), the worker has the option to either advance upstream beyond the immediate station to begin the next work unit earlier or proceed downstream beyond the current station border to complete the task on the current work unit. Regardless of the situation, there are typically practical constraints on the extent to which the worker can go in either direction, resulting in a situation where the worker must maintain a certain pace with a safety buffer. The terms upstream allowance and downstream allowance are occasionally employed to denote these boundaries in motion. Regardless of the specific methods used, as long as the worker consistently maintains a pace that aligns with the cycle time, the line will achieve the necessary cycle rate.

The next level of pacing is when there is no pacing, indicating that there is no specific deadline for completing the work at the station. Essentially, every assembly operator works at an individualized speed. This situation can arise under three conditions: (1) when manual labor is employed for transportation on the line, (2) when work units can be taken off the conveyor, allowing the worker to take as much time as needed to finish a specific unit, or (3) when an asynchronous conveyor is utilized and the worker has control over the release of each work unit from the station. In all of these instances, there is no mechanical method of establishing a pace discipline on the line. In order to meet the necessary production rate, the workers are incentivized to maintain a specific pace, either through their own shared work ethic or through a company-sponsored incentive structure.

4.2.5 Coping with Product Variety

Human workers are flexible, so manual assembly lines can be set up to handle a wide range of goods. Usually, the range of products must be fairly soft.

There are three distinct types of assembly line:

- Single model

- Batch model
- Mixed model

There is only one product that **a single-model line** makes in big quantities. Since each work unit is the same, the job that is done at each station is the same for all goods. This type of line is for items that are in high demand.

Batch-model and mixed-model lines are both made to make two or more goods or models, but they handle the differences between the models in different ways. Each product on **a batch-model line** is made in groups. Workstations are set up to make the necessary number of the first product. The stations are then moved around to make the next product, and so on. When demand for each product is average, they are often put together in groups. It makes more sense to use one assembly line to make a bunch of different goods at once than to build a separate line for each model.

Setting up the workstations means giving each station on the line a job to do, along with any special tools that are needed to do the job and the way the station is set up physically. Since most of the things made on the line are the same, the work that goes into making them is also the same. But because each model is different, jobs usually need to be done in a different order, and the tools needed at a given workstation for the last model might not be the same ones needed for the next model. The line may have to move more slowly because one model may take longer than another. For each new model, workers may need to be retrained or new tools may need to be bought. Because of this, changes need to be made to the way the station is set up before the next base model can be made. When these changes happen on a batch-model line, output time is lost.

On a **mixed-model line,** too, more than one model is made at the same time. Instead of being made in batches, each model is worked on at the same time. While one station is working on one model, the next station is working on a different model. Each station is equipped to do the different tasks needed to make any model that moves through it. Many consumer goods, like cars and big appliances, are put together on mixed-model lines. These products have different models, options, and sometimes even brand names.

A mixed-model line is better than a batch-model line because

- There is no downtime in production while switching between models.
- The excessive stockpiling that is common in batch production is avoided.
- The flexibility to adjust manufacturing rates of various models in response to variations in product demand. In contrast, a mixed-model line makes the issue of task assignment more complicated so that all workstations face the same workload. This sort of line makes scheduling (figuring out which models go where) and logistics (making sure the correct parts reach to each station for the model that's there) more challenging. Additionally, a batch-model line typically has more flexibility to handle a wider range of model configurations.

Figure 4.3 summarizes the subject by showing where the three types of assembly lines fall on a scale measuring product variety.

Product variety	Type of assembly line
Hard Variety	Batch –model line
Soft Variety	Mixed –model line
No Varietv	Single –model line

Figure 4.3 Three types of Manual Assembly Line Related to Product Variety

4.2.6 Study of single-model assembly lines

The correlations established in this section and the ones that follow are relevant to assembly lines with a single model.

4.2.6.1 Cycle Time and Workload Analysis

Planning the assembly line such that it achieves a production rate, S, high enough to meet product demand is essential. It is possible to convert the yearly quantity of product demand into an hourly rate. The number of shifts and hours worked by the line each week is a decision that must be made by management. Assuming the facility runs for 50 weeks a year, the necessary rate of production per hour is calculated by

$$S_p = \frac{R_a}{50SwHsh} \qquad (4.1)$$

Where Sp = average hourly production rate, units/hr:Ra = annual Requirement for the single product to be made on the line, units/yr: S_w = number of shifts/week; and H_{sh} = Hr/ shift. If the line operates 52 weeks rather than 50, then Sp = $Ra/52S_wH_{sh}$. The equation can be changed by using consistent time units if the product demand is measured over a period of time other than a year.

The cycle time T, the interval at which the line will be operated, must be converted from this production rate. A portion of manufacturing time will inevitably be lost due to unforeseen circumstances, such as power outages, equipment breakdowns, lack of a component required for assembly, quality issues, labor problems, and so on. It is important to factor this lost time into the cycle time. Because of these losses, the line efficiency measures how often the line is up and running comparative to the entire amount of time available throughout its shift. It is possible to calculate the cycle time by

$$T_c = \frac{60E}{S_p} \qquad (4.2)$$

where to cycle time of the line, min/cycle; S, required production rate, as determined from Equation (4.1), units per hour; the constant 60 converts the hourly production rate to a cycle time in minutes; and E = line efficiency. Typical values of E for a manual assembly line are in the range 0.90-0.98. The cycle time Tc establishes the ideal cycle rate for the line.

$$S_c = \frac{60}{T_c} \tag{4.3}$$

where S_c = cycle rate for the line, cycles/hr; and Tc is min/cycle as in Equation (4.2). This rate S must be greater than the required production rate S, because the line efficiency E is less than 100%. Line efficiency E is therefore defined as

$$E = \frac{S_p}{S_c} = \frac{T_c}{T_p} \tag{4.4}$$

where Tp = production cycle time (average value) (Tp = 60/Rp).

The construction of a finished object necessitates a specific cumulative duration. This is the **work content time** (T_{wc}), The entire time required to complete all work aspects necessary to produce one unit of product. It denotes the aggregate workload that the assembly line is responsible for completing on the product. Calculating the theoretical minimal number of workers needed on the assembly line to manufacture a product is beneficial with known T_{wc} and specified production rate S_p. Equation (4.5) in that section can be used to determine the number of workers on a production line:

$$w = \frac{WL}{AT} \tag{4.5}$$

where w is the number of employees in the queue; The variables WL and AT represent the workload to be completed in a given time period (minutes per hour) and available time per worker (minutes per hour/worker), respectively. The 60-minute time frame will be of interest. The product's work content time multiplied by the hourly production rate represents the workload during that time.

$$WL = S_p T_{wc} \tag{4.6}$$

Where S_p = rate of production, pc/hr; and T_{wc}= work content time, min/pc.

'Line efficiency is basically the same as availability, Availability is the more general reliability term; line efficiency is the term associated with production lines.

Equation (4.2) can be rearranged to the form S_p, = $60E/T_c$. Substituting this into

Equation (4.6),

$$WL = \frac{60ET_{wc}}{T_c}$$

Available time AT = one hour (60 min) multiplied by the proportion uptime on the line; that is,

AT = 60E

putting these terms for WL and AT into Equation (4.5), the equation reduces to the ratio T/T. Because the number of workers must be an integer,

$$w^* = \text{Minimum Integer} \geq 60E\,\frac{T_{wc}}{T_c} \tag{4.7}$$

where the minimum theoretical number of workers is w*. This ratio also indicates the theoretical minimum number of stations on the line if each workstation has a single worker. It is quite unlikely that this minimum theoretical value will be attained in real life. Two important elements that are present in actual assembly lines but are ignored in equation (4.7) tend to raise the 1520 d number of workers above the theoretical minimum:

- Due to repositioning losses, which occur at each station when workers or work units are moved, the total amount of time available for assembly is less than T.
- Another issue is the line balancing problem, which occurs when the amount of work that needs to be done at each station is not distributed properly, meaning that some stations will have less work than Te. As a result, the number of workers needed to complete the task increases.

4.2.6.2 Repositioning Losses

Repositioning losses occur in a production line when time is needed throughout each cycle to move the worker, the work unit, or both. For instance, in a continuous transportation system where work units are connected to a conveyor and traveling at a consistent velocity, it takes time for the worker to walk from the recently completed unit to the unit entering the station upstream. In alternative conveyor systems, a certain amount of time is necessary to extract the work unit from the conveyor and position it at the station where the worker can carry out their task on it. Repositioning in manual assembly lines inevitably results in some amount of lost time. T is defined as the duration needed for each cycle to relocate the worker, the work unit, or both. In the following study, we assume that the value of T is consistent for all workers, even though the real periods for repositioning may differ between stations.

To calculate the time available for the assembly activity at each workstation, the repositioning time T needs to be deducted from the cycle time T. The duration required to complete the given work at each station is referred to as the service time. The symbol Ti represents a station, where i is an identifier ranging from 1 to n. The variation in service times among stations is due to the uneven allocation of total work content. Certain stations will have a greater workload compared to others. There will be at least one station where T_si reaches its maximum value. The term "bottleneck station" is used to describe this particular station since it determines the amount of time it takes for the entire line to complete a cycle. The maximum service time should not exceed the difference between the cycle time T_c and the repositioning time T_r, expressed as:

$$Max\{T_{si}\} \leq T_c - T_r \quad \text{for } i = 1, 2, n \qquad (4.8)$$

Where $Max\{T_{si}\}$= maximum service time among all stations, min/cycle; T_c = cycle time for the assembly line form Equation (4.2), min/cycle; and T_r, = repositioning time (as- eferendo sumed the same for all stations), min/cycle. For simplicity of notation, let $T_s$denote this Under a maximum allowable service time:

$$T_s = Max\{T_{si}\} \leq T_c - T_r \qquad (4.9)$$

At all stations where $T_{si}$is less than T_s workers will be idle for a portion of the cycle, as portrayed in Figure 4.4. If the maximum service time does not consume the entire avail- able time (i.e., when $T_s < T_c - T_r$), it means that the line could be operated at a faster pace than T from Equation (4.2). In this case, the cycle time Te is usually reduced so that $T_c = T_s + T_r$ T+T; this allows production rate to be increased slightly.

The amount of time that can be spent on productive assembly work on the line is decreased by repositioning losses. One way to express these losses is as an efficiency factor:

$$E_r = \frac{T_s}{T_c} = \frac{T_c - T_r}{T_c} \quad (4.10)$$

where E_r = repositioning efficiency and the other terms are defined earlier.

4.2.7 Line Balancing Problem

The tasks carried out on an assembly line comprise numerous individual and different labor components. Typically, there are limitations on the order in which these aspects can be carried out, to varying degrees. Additionally, the line must operate at a specific rate of production, which can be converted into a necessary cycle time using Equation (4.2). Distribute these elements among workstations in a manner that ensures each worker receives an equitable workload. This part explores the vocabulary associated with the line balancing problem, while section 4.3 presents a variety of techniques to address this problem.

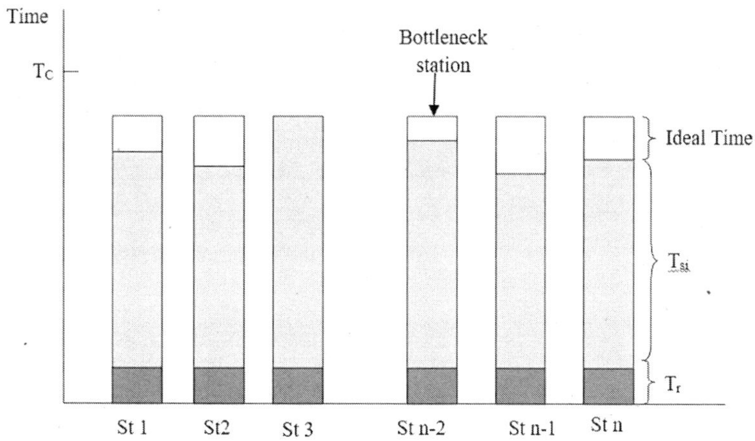

Figure 4.4 Components of cycle time at several stations on a manual assembly line. At the slowest station, the bottleneck station, idle time = zero; at other stations idle time exists

St.= workstation, n = number of workstations on the line, T_r = repositioning time, T_i = service time, Tc = cycle time.

Line balancing involves two crucial concepts: dividing the overall work content into the smallest logical work pieces and ensuring that these elements adhere to specific precedence requirements. Performance measurements can be established for solutions to the line balancing problem, using these concepts.

Minimum Rational Work Elements. A minimum rational work element refers to a little and specialized task that serves a restricted aim, such as adding a component to the base part, connecting two components, or completing a small piece of the overall work. A minimum reasonable work piece is indivisible without compromising practicality. For instance, the act of creating a hole in a sheet of metal or joining two machined components with a bolt or screw might be classified as minimum rational work elements. There is no logical need to break down these duties into smaller components of work. The sum of the product of the work element and time is equal to the product of the work content and time; that is,

$$T_{WC} = \sum_{k=1}^{nt} T_{ek} \tag{4.11}$$

where T_{ek} time to perform work element k, min; and n = number of work elements, that is, k = 1, 2..., n_e.

When it comes to line balancing, T_{ek} values are assumed to be constant and additive. However, this is not entirely accurate. The time it takes to complete work elements can vary, which introduces the issue of task time variability. Additionally, combining multiple work elements can sometimes result in motion economy, meaning that the time it takes to perform these elements in sequence may be less than the sum of their individual times. Despite these discrepancies, these assumptions are still made in order to simplify the process of solving the line balancing problem.

he task time at station i, or service time as it is sometimes written, T is made up of the work element times that have been given to that station.

$$T_{SI} = \sum_{k=1}^{n} T_{ek} \tag{4.12}$$

One underlying idea is that all T are less than the maximum service time T. The station service times are then broken down into logical jobs and given to workers. Ta Different parts of the work need different amounts of time, and those times aren't always the same. This means that because of the differences in work part times, some workers will be given more work and others will be given less.

Service times are different at each place, but they all have to add up to the work content time:

$$T_{WC} = \sum_{1=1}^{n} T_{si} \tag{4.13}$$

Precedence Constraints: Furthermore, the presence of fluctuating element times poses a challenge in achieving uniform service times throughout all stations. Additionally, there are limitations on the order of Precedence Constraints. Furthermore, there is a range of different time intervals in which the task elements can be executed. Certain elements must be completed prior to others.

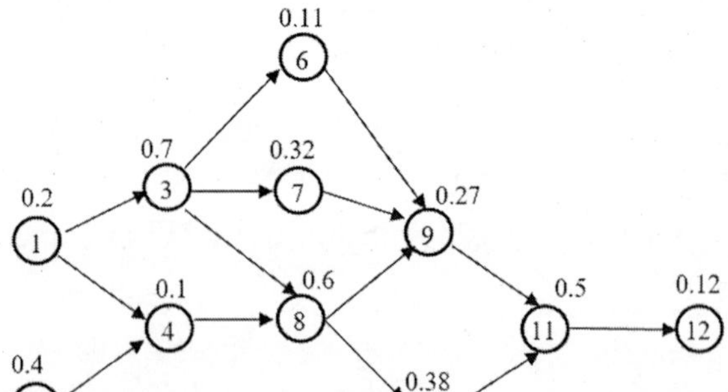

Figure 4.5 Precedence Diagram for Example 4.1

Arrows show the order in which the work elements must be completed, while nodes represent the individual work elements. Above every node are the element times.

To illustrate, in order to form a threaded hole, it is necessary to first drill a hole and then proceed to tap it. Prior to fastening a machine screw to join a matching component, it is necessary to first drill and tap the hole. The technological criteria that dictate the order in which tasks must be completed are referred to as precedence constraints. They make the line balancing situation more complex.

Precedence constraints can be visually represented using a precedence diagram, which is a network diagram that shows the order in which the task pieces need to be completed. Nodes represent work elements, and the connections between nodes define the precedence requirements. The series progresses in a linear manner from the left side to the right side. Figure 4.5 displays the precedence diagram for the given case, showcasing the terminology and a selection of the equations discussed in this context.

Example 4.1: A production line dedicated to a single model will be used to make a small electrical device. The product's assembly procedure has been reduced to just the particular steps listed in Table 4.4. The table lists the times for each step as well as the precise sequence in which they need to be completed. To fulfill a 100,000-unit annual demand, the line must be optimized. With five shifts per week that last 7.5 hours each, the line will run for 50 weeks a year. Each station will have one person assigned to it. Based on past experience, the line is expected to have an uptime efficiency of 96% and a repositioning time loss of 0.08 minutes per cycle. Calculate the total work content time (T_{WC}), required hourly production rate (RP), cycle time (T_C), theoretical minimum number of workers required on the line, and service time (T_S) needed to achieve the annual demand.

S. No.	Work Element Description	T_{ek} (min)	Must be Preceded By
1	Place a frame in work holder and clamp	0.2	-
2	Assemble plug, grommet to power cord	0.4	-
3	Assemble brackets to frame	0.7	1
4	Wire power cord to motor	0.1	1,2
5	Wire power cord to switch	0.3	2
6	Assemble mechanism plate to bracket	0.11	3
7	Assemble blade to bracket	0.32	3
8	Assemble motor to bracket	0.6	3,4
9	Align blade and attach to motor	0.27	6,7,8
10	Assemble switch to motor brackets	0.38	5,8
11	Attach cover, inspect, and test	0.5	9,10
12	Place in tote pane for packing	0.12	11

Table 4.3 Work Elements for Example 4.1

Solution:

(a) The total work element times in Table 4.4 add up to the total work content time.

$T_{WC} = 4.0$ min

(b) Based on the yearly demand, the rate of production per hour is

$$R_P = \frac{100{,}000}{50(5)(7.5)} = 53.33 \text{ units/hr}$$

(c) The corresponding cycle time T, with an uptime efficiency of 96% is

$$T_C = \frac{60(0.96)}{53.33} = 1.08 \text{ min.}$$

(d) Equation (4.7) provides the theoretical minimum number of workers.

$$W^* = \text{Min Int} \geq \frac{4.0}{1.08} = 3.7 \text{ rounded up to 4 workers}$$

(e) The available service time against which the line must be balanced is

$$T_S = 1.08 - 0.08 = 1.00 \text{ min}$$

4.2.8 Measures of Line Balance Efficiency

Due to variations in the minimum rational work element timings and the presence of precedence limitations among the components, achieving a flawless line balance is quite challenging. Criteria should be established to quantify the effectiveness of a particular line balancing solution. One such metric is the balancing efficiency, which is calculated by dividing the work content time by the total available service time on the line.

$$E_b = \frac{T_{WC}}{wT_S} \qquad (4.14)$$

E_b represents the balance efficiency, which is typically given as a percentage. T_S is the maximum available service time on the line, expressed as the maximum value of T_{SI} each cycle. Lastly, w represents the number of workers. The denominator in Equation (4.14) represents the entire amount of time that can be used on the line for assembling a single unit of the product. As the values of T_{WC} and wT_S go closer, the amount of idle time on the line decreases. E_b is a quantitative assessment of the quality of the line balancing solution. An optimal line balance results in a value of E_b 1.00. The line balancing efficiencies commonly seen in industry fall within the range of 0.90 to 0.95.

Balance delay is the opposite of balance efficiency and refers to the time lost as a result of imperfect balancing, expressed as a ratio to the total available time.

$$d = \frac{(wT_S - T_{wc})}{wT_S} \qquad (4.15)$$

where d = balance delay. A balance delay of zero indicates a perfect balance. Note that $E_b + d = 1$.

Requirements for Workers. When examining the connections in this section, three efficiency variables have been recognized that decrease the productivity of a human assembly line:

- Line efficiency, as defined by Equation (4.4), is the percentage of uptime on line E.
- The definition of repositioning efficiency, E_r, in Equation (4.10)
- Efficiency of line balancing, E_b, as given by Equation (4.14).

Together, they constitute the overall labor efficiency on the assembly line:

Assembly line labor efficiency = EE_rE_b (4.16)

Based on the previous Equation (4.7), a more practical estimate for the number of workers on the assembly line can be determined using this labor efficiency measure:

$$w = \text{Minimum Integer} \geq \frac{S_pT_{WC}}{60EE_rE_b} = \frac{T_{WC}}{E_rE_bT_C} = \frac{T_{WC}}{E_bT_S} \quad (4.17)$$

The variables in the equation are as follows: w represents the actual number of workers necessary on the line, Sp represents the hourly production rate in units per hour, and T_{wc} represents the work content time per product to be done on the line, measured in minutes per unit. The challenge with this relationship is in the complexity of establishing values for E, E_r, and E_b prior to the construction and operation of the line. However, the equation accurately represents the factors that influence the number of workers needed to do a specific amount of work on an assembly line with only one type.

4.2.9 Computerized assembly line balancing

Computerized assembly line balancing refers to the process of optimizing the allocation of tasks or workstations in an assembly line to maximize efficiency and productivity. It involves determining the sequence of operations, assigning tasks to workstations, and balancing the workload across these workstations to minimize idle time and maximize throughput.

There are several techniques and algorithms used for computerized assembly line balancing, including:

a) **Precedence Diagram Method (PDM):** This method involves creating a precedence diagram that illustrates the sequence of tasks and their dependencies. Algorithms like the ranked positional weight method and the longest task time method can then be applied to assign tasks to workstations and balance the line.

b) **Heuristic Algorithms**: These algorithms, such as the largest candidate rule, the positional weight rule, and the shortest processing time rule, provide quick and practical solutions for assembly line balancing problems by making iterative improvements to the initial line layout.

c) **Optimization Algorithms:** Techniques like genetic algorithms, simulated annealing, and ant colony optimization are used to find near-optimal solutions for complex assembly line balancing problems. These algorithms search the solution space to find the best arrangement of tasks and workstations based on defined optimization criteria.

d) **Simulation and Modeling**: Computerized simulation software allows engineers to model and analyze different assembly line configurations to evaluate their performance and identify opportunities for improvement before implementing changes in a real production environment.

By leveraging computerized assembly line balancing techniques, manufacturers can improve productivity, reduce cycle times, minimize costs, and enhance overall operational efficiency in their

Several software solutions are available for computerized assembly line balancing, each offering various features and capabilities to support different aspects of the process. production processes. Table 4.5 shows the list of software available to do line balancing. Detail of each software is beyond the scope of this book.

S. No.	Software
1	Arena
2	Flex Sim
3	Simul 8
4	Any Logic
5	Lanner witness
6	Opt. Quest
7	Factory CAD
8	Teamcenter

Table 4.4 List of software's Used for Line Balancing

Figure 4.6 Pictorial View of Team Center Line Balancing Software

4.3 Case study

The passenger car manufacturing company is currently facing a significant surge in demand for their product in the market. However, they are encountering difficulties in meeting this high market demand. In order to enhance the productivity of their product, the company must pinpoint the bottleneck area that is hindering their manufacturing process.

Identifying bottleneck area: A company that operates a model mix welding line. In order to meet the paint shop's requirements, the company must increase the productivity of the mix model weld line, assuming that two models are being manufactured concurrently. The current situation of Line 1 is as follows:

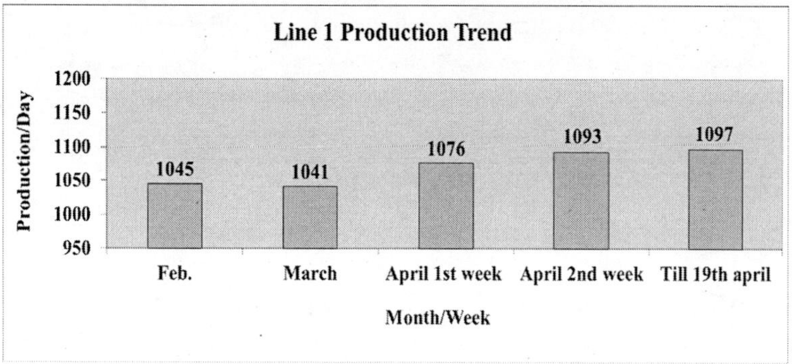

Figure 4.7 Line 1 Production Trend

Problem Identification: The company is required to produce 1128 vehicles per day, which entails operating the weld line at a 94 percent efficiency and a 66 second cycle time.

Approach:

1. Identify the bottleneck stations on weld line 1 by measuring cycle time of all sub assembly lines as showing in figure 4.8

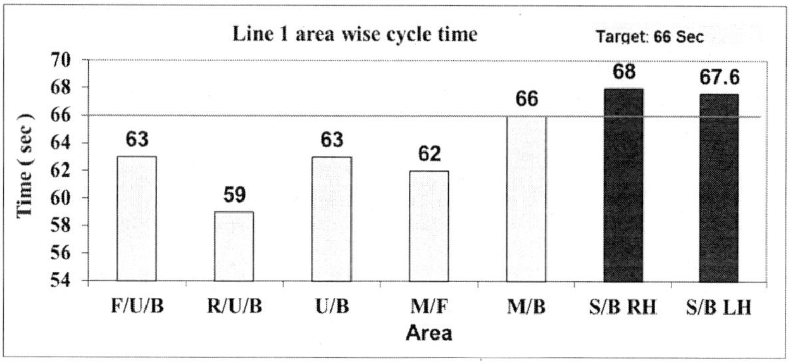

Figure 4.8 The primary bottleneck stations in the right-side body (S/B RH) and left side body (S/B LH) will be debottlenecked by redistributing 13 locations in S/B LH and 12 spots in S/B RH.

2. Implemented schedule for daily preventive maintenance work during break time in A and B shift.
3. Reduction in frequently occurring faults (602C hand clamp fault, side sill unit fault, roof shifting problem, AHC fault, OVC alarm fault) by taking proper countermeasure along with maintenance and production by following preventive maintenance schedule.
4. Implementation of proper format of "Daily log book" for keeping record of production related issues.
5. Measure the cycle time of all subassemblies after adopting the 5 recommendations that were presented.

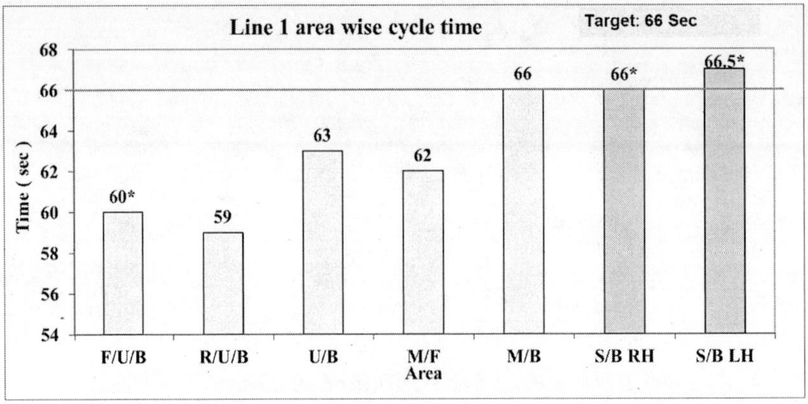

Figure 4.9 Line 1 Area Wise Cycle Time After Improvement

By applying all of these strategies, the company has achieved a production capacity of 1125 automobiles per day without requiring any further capital expenditure i.e. Company is able to provide more vehicles in market to meet market demand.

UNIT SUMMARY

Within this unit, we discussed about assembly lines in great detail, covering many important topics that are useful in engineering. The unit starts by going over the basics of assembly lines. Then it moves on to assembly line balancing, which explains how to make the best use of job allocation and output methods. Ideas like splitting jobs, flexible and U-shaped line layouts, and mixed model line balance are talked about in detail, along with how they can be used and what benefits they offer. The unit also looks at what people think about assembly lines now, taking into account new ideas and progress in the field.

EXERCISES NO: 4

Multiple Choice Questions

1. Which of the following is a characteristic of assembly lines?
    a. Randomized workflow
    b. Sequential production
    c. Irregular task distribution
    d. Inconsistent product output
2. What is the primary objective of assembly line balancing?
    a. Maximizing downtime
    b. Minimizing task variation
    c. Increasing worker fatigue
    d. Reducing production efficiency
3. What is the process of dividing tasks among workers to optimize efficiency called?
    a. Task isolation

b. Task delegation
c. Task allocation
d. Task amalgamation

4. Which type of line layout allows for greater adaptability and customization of production processes?
a. Linear layout
b. Circular layout
c. Flexible layout
d. Static layout

5. What is the term for balancing assembly lines that produce different models of products simultaneously?
a. Single model line balancing
b. Mixed model line balancing
c. Uniform model line balancing
d. Sequential model line balancing

6. What reflects current thoughts on assembly lines regarding efficiency and productivity?
a. Emphasizing manual labor over automation
b. Focusing solely on speed rather than quality
c. Incorporating lean manufacturing principles
d. Ignoring employee feedback and suggestions

7. Which method utilizes computer algorithms to optimize task allocation and sequencing in assembly line production?
a. Manual assembly line balancing
b. Randomized task assignment
c. Computerized assembly line balancing
d. Intuitive task distribution

8. What is the primary function of an assembly workstation?
a. Testing products
b. Storing finished goods
c. Assembling components
d. Conducting market research

9. Which factor is NOT considered when designing assembly workstations?
a. Worker safety
b. Efficiency of assembly process
c. Proximity to cafeteria
d. Ergonomics

10. Which type of assembly workstation layout promotes efficient material flow and minimizes unnecessary movement?

a. Random layout
b. U-shaped layout
c. Circular layout
d. Linear layout

11. What is a characteristic of a well-designed assembly workstation?
    a. Limited space for movement
    b. Cluttered layout
    c. Poor lighting conditions
    d. Ergonomic design
12. In a just-in-time manufacturing environment, assembly workstations are designed to:
    a. Maximize inventory storage
    b. Minimize worker efficiency
    c. Reduce setup time and lead times
    d. Increase batch production
13. Which type of assembly workstation is designed for repetitive tasks performed by a single operator?
    a. Group workstation
    b. Flexible workstation
    c. Individual workstation
    d. Collaborative workstation
14. Which assembly workstation design allows for simultaneous assembly of different components by multiple workers?
    a. Group workstation
    b. Cellular workstation
    c. Individual workstation
    d. Sequential workstation
15. What factor is crucial for ensuring worker safety in assembly workstations?
    a. Lack of safety equipment
    b. High noise levels
    c. Poor ventilation
    d. Proper training and supervision
16. What does "cycle time" refer to in manufacturing?
    a. The time taken to complete a task
    b. The time between the start of two successive tasks
    c. The time taken for machine breakdown
    d. The time taken to train a worker
17. Which of the following best defines "takt time"?
    a. The time needed to repair a machine

b. The time taken to set up a production line
c. The maximum time allowed to produce one unit to meet customer demand
d. The time taken for maintenance checks

18. In a manufacturing process, if the cycle time is greater than the takt time, what can be inferred?
a. The process is running efficiently
b. The process is operating at a slower pace than required to meet customer demand
c. The process is overproducing
d. The process is underutilizing resources

19. How is cycle time typically measured?
a. In hours
b. In minutes
c. In days
d. In weeks

20. Which of the following statements best describes the purpose of takt time?
a. To minimize production costs
b. To maximize machine utilization
c. To synchronize production with customer demand
d. To reduce cycle time

21. If a manufacturing process has a takt time of 5 minutes and a cycle time of 6 minutes, what action should be taken?
a. Increase production speed
b. Decrease production speed
c. Maintain current production speed
d. Stop production temporarily

22. Which term represents the heartbeat of production, indicating the rate at which products need to be produced to satisfy customer demand?
a. Cycle time
b. Takt time
c. Lead time
d. Downtime

23. What happens if the cycle time is less than the takt time in a production process?
a. The process is overproducing
b. The process is running at optimal efficiency
c. The process is under producing
d. The process is not properly synchronized with customer demand

24. How does understanding cycle time and takt time help in improving production efficiency?
a. By reducing worker breaks
b. By increasing machine downtime
c. By identifying bottlenecks and optimizing processes

d. By extending lead times

25. Which approach is commonly used to balance the product mix in assembly lines?

a. Random allocation of products to workstations
b. Assigning products with similar processing times to the same workstation
c. Prioritizing high-demand products over low-demand ones
d. Ignoring product mix altogether

Answers of Multiple-Choice Questions

1 b, 2 b, 3 c, 4 c, 5 b, 6 c, 7 c, 8 c, 9 c, 10 b, 11 d, 12 c, 13 c, 14 a, 15 d, 16 a, 17 c, 18 b, 19 b, 20 c,21 b,22 b,23 a,24 c,25 b.

Short and Long Answer Type Questions

1. What do you mean by 'Assembly? What are the characteristics of assembly process?
2. What are the various methods of Assembly process? Explain briefly.
3. Differentiate between Manual and Automated Assembly.
4. What do you mean by Assembly system?
5. Explain single point Manual Assembly.
6. What is line assembly? Explain the features of Manual line Assembly.
7. What do you mean by line balancing? Why it is necessary?
8. Describe the process of task splitting in assembly line operations. How does it contribute to overall efficiency?
9. Explore the role of computerized systems in assembly line balancing. What are the benefits and limitations of relying on such technology?

KNOW MORE

REFERENCES

[1] Andreasen, M., S., Kahler, and T., Lund, Design for Assembly, IFS (Publications) Ltd., UK, and Springer-Verlag, Berlin, 1983.

[2] BARD, J. F., E. M. DAR-EL, and A. SHTUB, & quot; An Analytical Framework for Sequencing Mixed-Model Assembly Lines, & quot; International Journal of Production Research, Vol. 30, No. 1, 1992, pp. 35-48.

[3] BLACK, J. T., The Design of the Factory with a Future, McGraw-Hill, Inc., New York, NY, 1991.

[4] *BOOTHROYD, G., P. DEWHURST, and W. KNIGHT, Product Design for Manufacture and Assembly, Marcel Dekker, Inc., New York, 1994.*

[5] *BRALLA, J. G., Handbook of Product Design for Manufacturing, McGraw-Hill Book Company, New York, 1986, Chapter 7.*

[6] *CHOW, W.-M., Assembly Line Design: Methodology and Applications, Marcel Dekker, Inc., New York, 1990.*

[7] *DAR-EL, E. M., & quot.; Mixed-Model Assembly Line Sequencing Problems, & quot; OMEGA, Vol. 6. No. 4, 1978, pp. 313-323*

[8] *DAR-EL, E. M., and A. NAVIDI, & quot.; A Mixed-Model Sequencing Application, & quot; International Journal of Production Research, Vol. 19, No. 1, 1981, pp. 69-84.*

[9] *DAR-EL, E. M., and F. COTHER, & quot.; Assembly Line Sequencing for Model Mix, & quot; International Journal of Production Research, Vol. 13, No. 5, 1975, pp. 463–477.*

[10] *DEUTSCH, D. F., & quot.; A Branch and Bound Technique for Mixed-Model Assembly Line Balancing, & quot.; Ph.D. Dissertation (Unpublished), Arizona State University, 1971.*

[11] *FERNANDES, C. J. L., & quot.; Heuristic Methods for Mixed-Model Assembly Line Balancing and Sequencing, & quot.; M.S. Thesis (Unpublished), Lehigh University, 1992.*

[12] *FERNANDES, C. J. L., and M. P. Groover, & quot.; Mixed-model Assembly Line Balancing and Sequencing: A Survey, & quo.t; Engineering Design and Automation, Vol. 1, No. 1, 1995, pp. 33-42.*

[13] *HELGESON, W. B., and D. P. BIRNIE, & quot.; Assembly Line Balancing Using Ranked Positional Weight Technique, & quot.; Journal of Industrial Engineering, Vol. 12, No. 6, 1961, pp. 394-398. [14]*

[14] *HOFFMAN, T., & quot.; Assembly Line Balancing: A Set of Challenging Problems, & quot; International Journal of production Research, Vol. 28, No. 10, 1990, pp. 1807-1815.*

[15] *IGNALL, E. J., & quot; A Review of Assembly Line Balancing, & quot; Journal of Industrial Engineering, Vol. 16, No. 4, 1965, pp. 244-252.*

[16] *KILBRIDGE, M., and WESTER, L., & quot.; A Heuristic Method of Assembly Line Balancing, & quot; Journal of Industrial Engineering, Vol. 12, No. 6, 1961, pp. 292–298.*

[17] *MACASKILL, J. L. C., & quot.; Production Line Balances for Mixed-Model Lines, & quot; Management Science, Vol. 19, No. 4, 1972, pp. 423-434.*

[18] *MOODIE, C. L., and YOUNG, H. H., & quot; A Heuristic Method of Assembly Line Balancing for assumptions of Constant or Variable Work Element Times, & quot; Journal of Industrial Engineering, Vol. 16, No. 1, 1965, pp. 23-29.*

[19] *NOF, S. Y., W. E. WILHELM, and H.-J. WARNECKE, Industrial Assembly, Chapman & Hall, how London, UK, 1997.*

[20] *PRENTING, T. O., and N. T. THOMOPOULOS, Humanism and Technology in Assembly Systems, Hayden Book Company, Inc., Rochelle Park, NJ, 1974.*

[21] *REKIEK, B., and A. DELCHAMBRE, Assembly Line Design: The Balancing of Mixed-Model*

---------*--------*---------

5 MATERIAL MANAGEMENT

UNIT SPECIFICS

The following topics are covered in this unit:

- Material management
- Purchasing and Procurement
- Stores and Storekeeping
- Inventory and Inventory control
- Economic order Quantity
- Vendor Development

All the topics are discussed in a lucid manner along with their engineering applications. For student's practice, a moderate number of multiple-choice questions, short and long answer type questions and numerical problems are also given. There are further suggested readings and reference for deep learners and reader's assistance.

RATIONALE

In this unit, a comprehensive exploration of material management is provided, covering various essential aspects crucial to engineering applications. Beginning with the fundamentals of material management, the unit delves into Purchasing and procurement, functions of store and store keeping. Concepts such as tender processing, purchase process, and quotations analysis are thoroughly discussed, offering insights into their practical implementation and benefits. Additionally, the unit delves into inventory control, Vendor development and Economic order quantity in the industry. Moreover, students are provided with ample opportunities for practice through a diverse range of exercises, including multiple-choice questions, short and long answeı type questions, and numerical problems, ensuring a comprehensive understanding and mastery of the subject matter. In this chapter all such areas are covered in details.

PRE-REQUISITES

Fundamental Knowledge of any Manufacturing Industry Layout

UNIT OUTCOMES

After completing this unit the reader will be able to:

U5-O1: Understand the managerial concept of Material Supply in Industry

U5-O2: Understand the Procurement process of Material.

U5-O3: Understand the importance of store and prime function of store keeper

U5-O4: Understand the Concept of Inventory and Economic order Quantity.

U5-O5: Understand the role of Vendor in Manufacturing Industry.

Unit-5 Outcomes	EXPECTED MAPPING WITH COURSE OUTCOMES (1- Weak Correlation; 2- Medium correlation; 3- Strong Correlation)				
	CO-1	CO-2	CO-3	CO-4	CO-5
U5-O1	3	1	2	2	3
U5-O2	2	0	1	2	3
U5-O3	3	1	2	2	3
U5-O4	2	1	2	2	3
U5-O5	3	1	1	3	3

5.1 Material management

5.1.1 Introduction

Material Management aims for integrated approach towards the management of material. Manufacturing industries spend approximately 58% of the money for managing and procuring materials. i.e., material involves a substantial portion of the capital invested in an industrial concern.

It means a small saving in material Handling, storing and procurement impact a lot on production cost. Material Management includes managing the type, amount, location, movement, timing of purchase of various materials etc., used in an industrial concern.

Material management is an important aspect of supply chain management that involves the planning, storage procurement, control, and distribution of materials to guarantee the efficient use of resources within an organization. It plays a pivotal role in attaining the organization's goals by optimizing the flow of materials and confirming that the right quantity of materials is available at the right place and time.

5.1.2 Functions of Material Management

1. Material Procurement Planning
    - Planning for KD (Knock Down) parts i.e. Overseas Vendors
    - Planning for Local Vendor Parts
2. Purchasing of Material
3. Receiving and Warehousing (Store Management)
4. Inventory Control
5. Standardizing and Value analysis
6. Transportation

- Minimize Truck Travel Time (TTT) for external Supply
- Minimize in house material Movement by effective Material handling techniques

7. Obsolescence of Material
   - Minimize the inventory of discontinue Models/Products.
8. Disposal of Scrap

5.1.3 Importance of Material Management

A survey conducted by the directorate of industrial statistics (1954-57) states that average material cost is 64 to 70 % of the sales value i.e. only 36 to 30 % cost is for wages, salaries, overheads and profit.

Material management is of paramount importance in the overall functioning and achievement of an organization. It plays a crucial role in various aspects of business operations, contributing to efficiency, cost-effectiveness, and customer satisfaction. These are key reasons highlighting the importance of material management:

1. **Cost Reduction:**
   - Efficient material management helps in minimizing costs associated with procurement, storage, and distribution.
   - Optimizing inventory levels prevents overstocking or stockouts, reducing carrying costs and potential losses.
2. **Operational Efficiency:**
   - Proper planning and coordination in material management streamline production processes, reducing lead times and enhancing overall operational efficiency.
   - Timely availability of materials ensures uninterrupted production and delivery schedules.
3. **Resource Optimization:**
   - Material management ensures the optimal use of resources by avoiding excess inventory and reducing wastage.
   - It helps in identifying the right quantity and quality of materials required for production.
4. **Improved Cash Flow:**
   - Effective inventory control prevents tying up excess capital in unsold or slow-moving inventory.
   - Efficient material management contributes to a positive cash flow by aligning procurement with actual demand.
5. **Enhanced Customer Satisfaction:**
   - Timely delivery of products is facilitated by efficient material management, leading to increased customer satisfaction.
   - Consistent product availability and reliability contribute to positive customer experiences.

6. **Strategic Advantage:**
   - ➢ Organizations with effective material management practices gain a competitive edge by responding quickly to market demands and changes.
   - ➢ Strategic sourcing and efficient logistics contribute to a more agile and approachable supply chain.
7. **Quality Control:**
   - ➢ Material management includes quality control measures to ensure that only high-quality materials are used in production.
   - ➢ Maintaining quality standards contributes to the production of reliable and durable products.
8. **Risk Management:**
   - ➢ Material management helps in recognizing and reducing the chances of supply chain disruptions, ensuring business continuity.
   - ➢ Diversification of suppliers and planned sourcing reduce dependency on a single source.
9. **Regulatory Compliance:**
   - ➢ Compliance with regulations related to inventory management, environmental concerns, and product safety is facilitated by effective material management practices.
10. **Information Visibility:**
    - ➢ Information systems and technology used in material management provide real-time visibility into register levels, demand forecasts, and supplier performance.
    - ➢ Data-driven decision-making becomes more accurate and responsive.

5.1.4 Objectives of Material Management

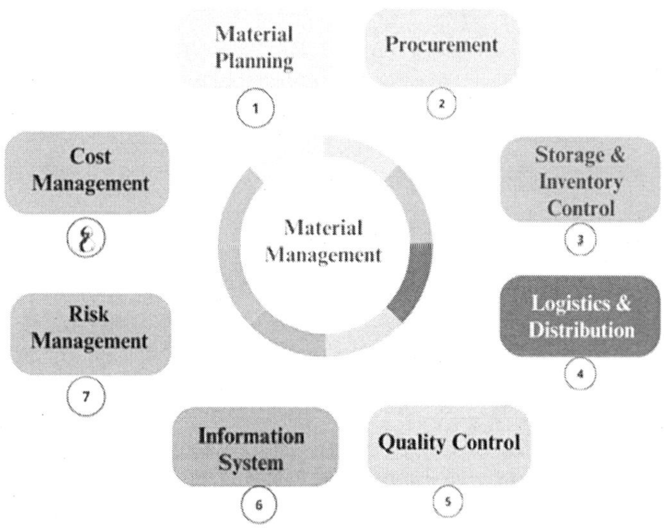

Figure 5.1: Objectives of Material Management

Effective material management contributes not only to the operational efficiency of a body but also to its competitiveness in the marketplace. Organizations that successfully implement sound material management practices can achieve cost savings, improve client satisfaction, and maintain a strategic advantage in the dynamic business environment

5.1.5 Role of Material Manager

DO necessary coordination with Production Department, Sales department and Supplier Quality assurance Department as shown in figure 5.2.

Figure 5.2: Role of Material Manager

5.2 Purchase and Procurement

5.2.1 Introduction

The purchase department plays an important role in the overall functioning of an industry, ensuring the efficient attainment of goods and services necessary for the organization's operations. This department is responsible for managing the entire procurement process, from recognizing the need for goods and services to the final receipt and payment. The prime objective is to obtain quality products and services at the most competitive price, within the specified timeframe.

5.2.2 Objectives of Purchase Department

- To procure right material
- To procure material in right quantity
- To procure material in right quality
- To develop Multiple reliable Vendors
- To procure material economically
- To receive and deliver materials at right place
- To receive and deliver the material at right time.

5.2.3 Functions of Purchase Department

The functions of the purchase department encompass various activities aimed at ensuring an efficient and cost-effective procurement process. Here are the key functions of a purchase department.

Needs Assessment:

- Identify and analyze the organization's requirements for goods and services.
- Collaborate with different departments to understand their needs and specifications.

Vendor Identification and Evaluation:

- Identify potential suppliers, vendors, and contractors.
- Evaluate and assess suppliers based on criteria such as quality, cost, reliability, and past performance.

Sourcing Strategies:

- Develop and implement sourcing strategies to secure the best value for the organization.
- Explore alternative suppliers and negotiate favorable terms and conditions.

Request for Proposals (RFPs) and Quotations:

- Issue RFPs or requests for quotations to potential suppliers.
- Review and analyze proposals, quotations, and bids to make informed decisions.

Negotiation:

- Negotiate terms and conditions with suppliers, including pricing, payment terms, delivery schedules, and quality standards.
- Ensure that contracts are favorable and aligned with organizational goals.

Purchase Orders:

- Generate and issue purchase orders once terms are agreed upon.
- Clearly outline specifications, quantities, delivery dates, and other relevant details.

Supplier Relationship Management:

- Establish and maintain positive relationships with suppliers.
- Monitor supplier performance, address issues promptly, and foster collaboration.

Quality Assurance:

- Implement quality control measures to ensure procured goods and services meet specified standards.
- Conduct inspections and quality checks as necessary.

Logistics and Inventory Management:

- Coordinate the logistics of receiving and storing purchased items.
- Manage inventory levels to prevent shortages or excess stock.

Cost Management:

- Monitor and control costs associated with procurement activities.
- Identify cost-saving opportunities and implement strategies to optimize spending.

Compliance and Risk Management:

- Ensure compliance with legal and governing requirements in procurement processes.
- Identify and manage hazards associated with procurement activities.

Technology Integration:

- Utilize procurement software and technology for efficient and streamlined processes.
- Implement e-procurement systems to facilitate electronic transactions and documentation.

Market Research:

- Stay informed about market movements, pricing, and new products or services.
- Conduct market research to identify potential suppliers and assess market conditions.

Contract Management:

- Manage contracts with suppliers, ensuring that both parties fulfill their requirements.
- Renew, renegotiate, or terminate contracts as needed.

Continuous Improvement:

- Regularly assess and improve procurement processes.
- Seek feedback from internal stakeholders and suppliers to enhance efficiency and effectiveness.

5.3 Organization structure of Procurement Department

Keep in mind that the actual structure may include more roles, additional levels of hierarchy, or different titles as per the organization's needs. This is a basic representation to give you an idea of how a purchase department staff structure might be organized

Model 1

- The "Head of Purchase Department" oversees the entire department.
- The "Procurement Manager" is responsible for managing procurement strategies and processes.
- "Buyers" are responsible for specific categories of goods or services and report to the Procurement Manager.
- "Assistant Buyers" support the Buyers in their day-to-day tasks.
- The "Vendor Relations Coordinator" focuses on building and maintaining positive relationships with suppliers.
- The "Logistics Coordinator" is responsible for coordinating the logistics of receiving and storing purchased items.

Model 2

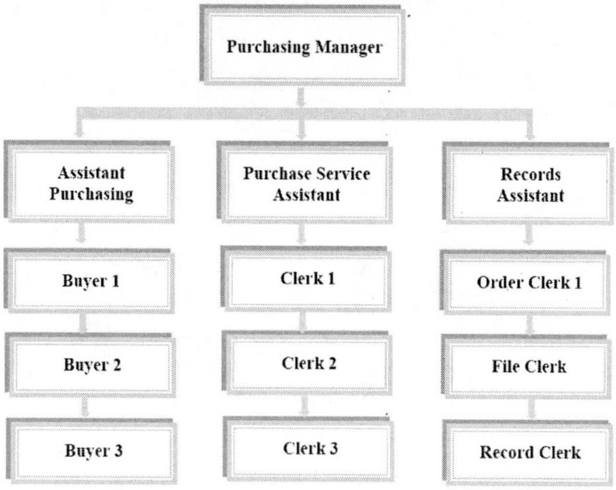

Figure 5.3: Organization Structure of Procurement Department

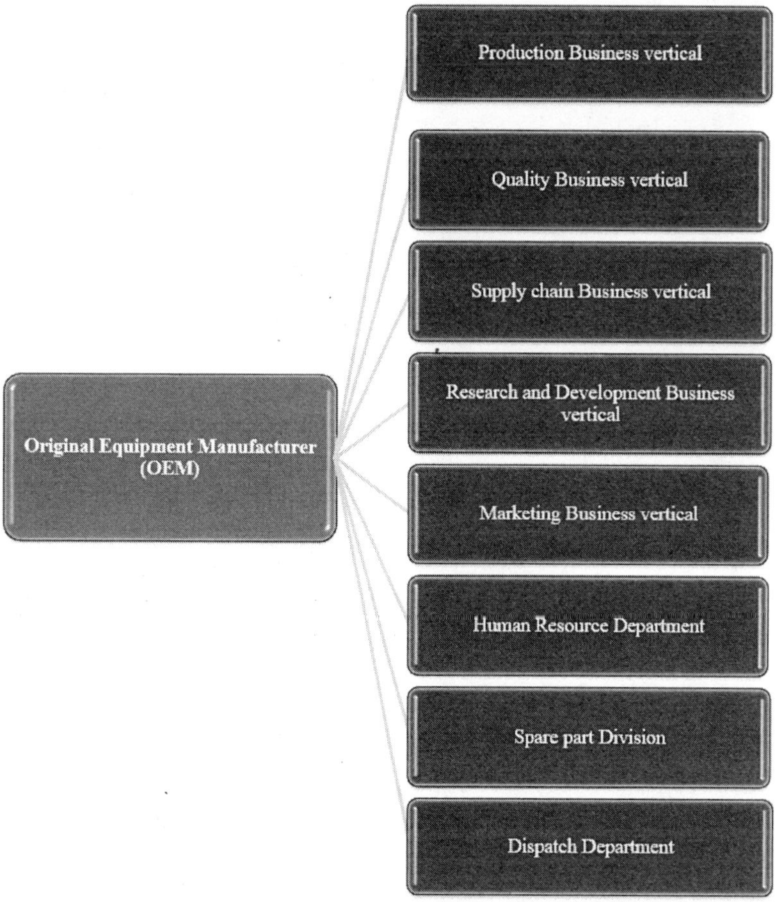

Figure 5.4: Organizational Structure

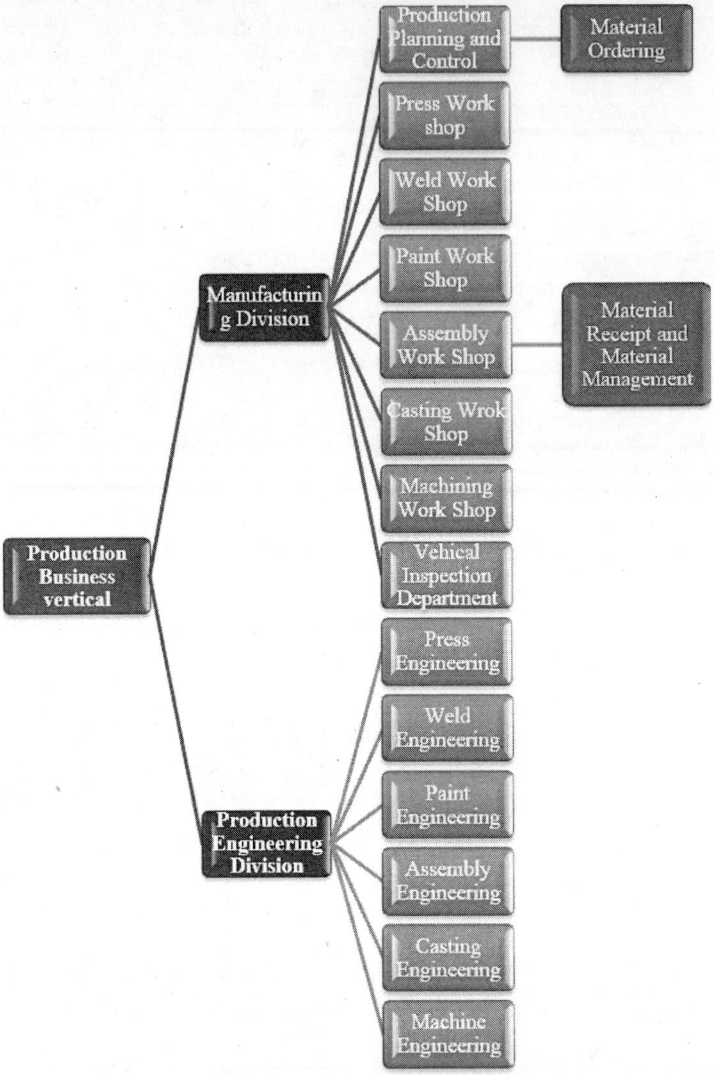

Figure 5.5: Organizational Structure of Production Business Vertical

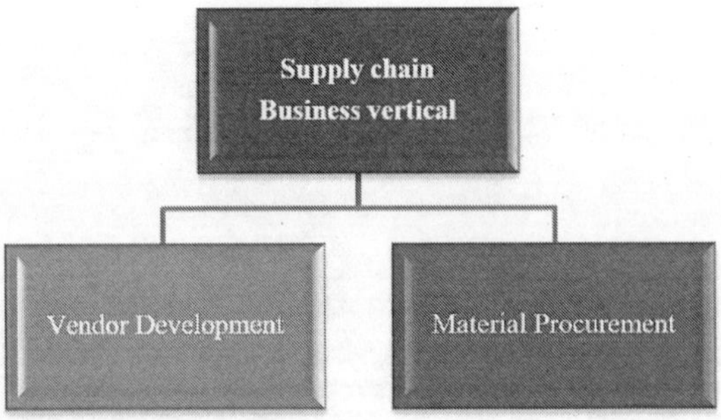

Figure 5.6: Organizational Structure of Supply Chain Business Vertical

Note: Material management department plays a pivotal role among all production and supply chain business vertical as shown in figure 5.4 (a),(b), (c).

5.3.1 Centralized and Decentralized Purchase Organization

Centralized and decentralized purchase organizations represent two different approaches to how procurement activities are structured within an organization. These approaches have inferences for decision-making, efficiency, and coordination. Let's explore the features of both centralized and decentralized purchase organizations

5.3.1.1 Centralized Purchase Organization

Advantages of Centralized Purchasing

Decision-Making:

- Decision-making authority is concentrated at a central point, usually within a dedicated purchasing or procurement department.
- High-level decisions, such as vendor selection and contract negotiation, are typically made centrally.

Coordination:

- Centralized coordination allows for standardized processes and uniform purchasing policies across the entire organization.
- There is a sole point of contact for vendors, streamlining communication and negotiations.

Efficiency:

- Centralization often leads to economies of scale, as bulk purchases and standardized processes can result in cost savings.
- Centralized control can facilitate better monitoring and control of expenditures.

Specialization:

- Specialized procurement professionals are often concentrated within the centralized purchasing department.
- Expertise can be leveraged for strategic sourcing and negotiations.

Consistency:

- Standardization of procurement practices ensures consistency in supplier relationships, contract terms, and quality standards.

Risk Management:

- Risks associated with procurement, such as compliance and contract management, can be more effectively managed through a centralized approach.

Disadvantages of Centralized Purchasing

- Centralized purchasing may lack the flexibility to cater to the unique needs and requirements of individual departments or business units within the organization.

- Some departments may feel restricted in their ability to make quick, independent purchasing decisions.
- Centralized decision-making processes can be slower, especially when departments need rapid responses to changing market conditions or unforeseen circumstances.
- The hierarchical approval process may contribute to delays in procurement activities.

Application of Centralized Purchasing

- A large hospital network centralizes its purchasing functions to acquire medical supplies, equipment, and pharmaceuticals. This allows the body to negotiate bulk purchases, ensuring a steady supply of high-quality medical products at lower costs.
- A retail chain with multiple branches centralizes its purchasing to procure inventory, merchandise, and other goods. By negotiating contracts centrally, the organization can achieve better pricing from suppliers and maintain a standardized product range across all stores.
- For purchasing small items of fairly high value such as tool bits, grinding wheels, gauges etc.

5.3.1.2 Decentralized Purchase Organization

Decision-Making

- Decision-making authority is distributed across various departments or business units within the organization.
- Individual departments or units have autonomy in making purchasing decisions based on their specific needs.

Coordination

- Each department or business unit handles its own purchasing activities, leading to decentralized coordination.
- Communication with vendors may occur independently in different parts of the organization.

Flexibility

- Decentralization allows for greater flexibility in responding to specific needs of different departments.
- Departments can tailor their procurement strategies to meet unique requirements.

Responsiveness

- Rapid response to changing market conditions or specific departmental needs is a key advantage of decentralized purchasing.
- Localized decision-making can lead to quicker adaptation to market dynamics.

Expertise

- Each division may have its own procurement professionals, possessing specialized knowledge related to their specific products or services.
- Expertise is distributed across the organization.

Innovation

Decentralized structures may encourage innovation in procurement practices as individual departments explore unique solutions.

Disadvantages of Decentralized Purchasing

- Involve duplication of work
- Less Quantity discounts
- Decentralized purchasing may miss out on probable cost savings that can be achieved through bulk purchasing and centralized negotiations with suppliers.

Application of Decentralized Purchasing

- Where different plants of a large organization require quite different types of materials.
- Where Purchase is to be made with in the local community to promote better public relationship.

The choice between centralized and decentralized purchase organizations often depends on factors such as the organization's size, industry, geographical dispersion, and the nature of its products or services. It's essential to weigh the advantages and disadvantages of each approach based on the organization's specific goals and operational requirements

In some organizations, a hybrid approach is adopted, combining elements of both centralized and decentralized structures. This allows for a balance between standardized processes and flexibility to meet specific needs at different levels within the organization.

5.4 Buying Procedures

The choice of technique often depends on features such as the nature of the material, the organization's requirements, market conditions, and strategic objectives. Here are some common techniques for buying or purchasing materials:

5.4.1 Spot Quotations

Spot quotation is a buying technique where a purchaser seeks immediate price quotations or bids from suppliers for goods or services. This method is typically employed when the purchaser has an urgent need or when the market situations are such that obtaining quotes quickly is essential. Process of spot quotation buying is shown in figure 5.4.

- The organization identifies an immediate need for specific goods or services.
- The purchaser quickly identifies potential suppliers capable of fulfilling the urgent requirement
- The purchaser issues a request for spot quotations to the selected suppliers.
- Suppliers respond promptly with their price quotations, terms, and conditions.
- The purchaser evaluates the received quotations based on factors such as price, delivery time, quality, and any other relevant criteria.
- If necessary, negotiations may take place to finalize terms with the selected supplier.
- Once a suitable supplier is identified and terms are agreed upon, the purchaser places the order.
- Suppliers are expected to fulfill the order promptly to meet the urgent requirement.

Advantages of Spot Quotation Buying

- Spot quotation buying allows organizations to quickly obtain prices and make procurement decisions.
- Offers flexibility in terms of supplier selection and negotiation.
- Suitable for situations where short-term cost considerations take precedence over long-term contracts.
- Encourages a responsive and agile association with suppliers who can quickly provide quotes and fulfill orders.

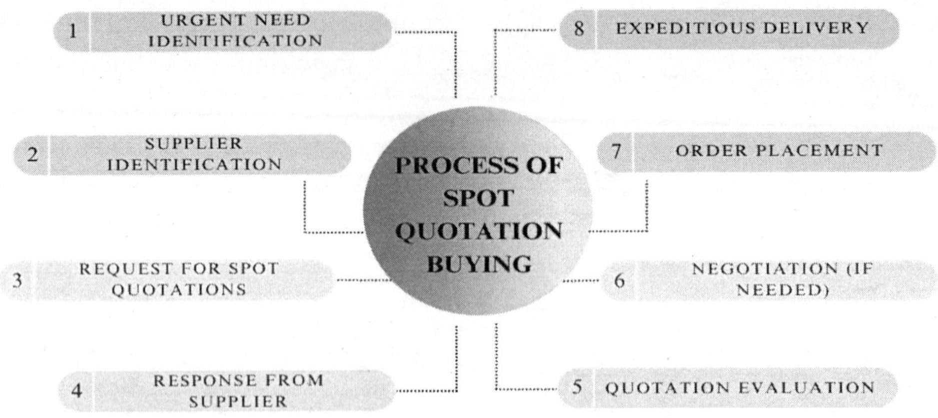

Figure 5.7: Process of Spot Quotation Buying

Challenges of Spot Quotation Buying:

- Relies on existing or readily available suppliers, limiting the pool of potential vendors.
- In urgent situations, suppliers may charge higher prices due to the immediate need.
- Spot quotation buying may lack the comprehensive contractual agreements often associated with longer-term procurement methods.
- Due to the urgency, there may be limited time for thorough quality assurance and supplier vetting.

The spot quotation buying technique is suitable for specific scenarios where speed and flexibility are critical.

5.4.2 Floating a Limited Enquiry

Floating a limited enquiry is a procurement process where an organization issues a request for information or quotation to a select group of pre-qualified suppliers or vendors. This approach is used when the organization wants to gather specific information or quotes for a particular requirement from a targeted pool of suppliers. A process is shown in figure 5.5.

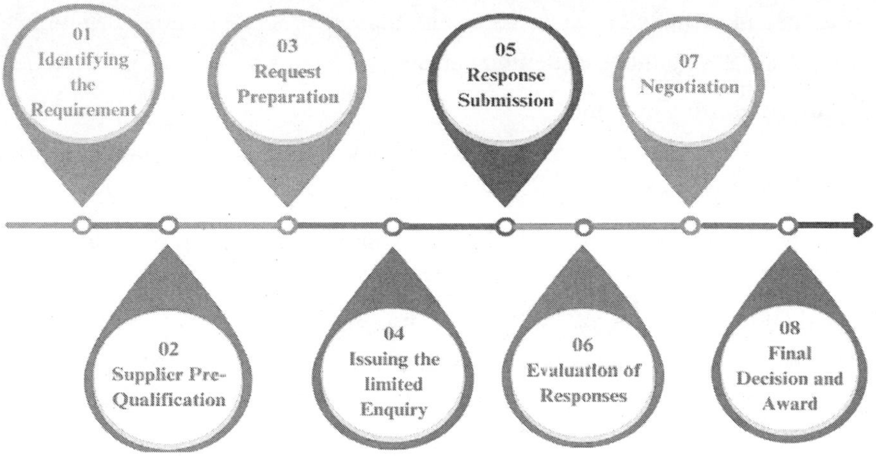

Figure 5.8: Process of Floating a Limited Enquiry Buying

- The organization identifies a specific procurement need or project for which it requires information or price quotations.
- The organization pre-qualifies a limited number of suppliers based on criteria such as capabilities, experience, financial stability, and compliance with relevant standards.
- The procurement team prepares a request for information (RFI) or request for quotation (RFQ) document outlining the details of the requirement, specifications, delivery terms, and any other relevant information
- The limited enquiry is sent to the pre-qualified suppliers, informing them of the specific procurement need and inviting them to submit their responses.
- Suppliers respond to the limited enquiry by providing the requested information or submitting price quotations
- The procurement team evaluates the responses based on criteria such as cost, quality, delivery time, and compliance with specifications.
- Based on the evaluation and any negotiations, the organization makes a final decision on the supplier to be awarded the contract.

Advantages of Spot Quotation Buying

- This approach is more efficient than a broad procurement process, as it targets a specific group of suppliers.
- The organization can gather targeted information or quotes for a particular requirement without involving a large number of suppliers.
- Limited enquiries are suitable for smaller or routine procurement needs, reducing the workload on both the organization and suppliers.
- The process is generally faster than open market or unrestricted procurement, making it suitable for urgent requirements.

- The procurement team can focus its evaluation efforts on a smaller number of supplier responses, facilitating a more in-depth analysis.

Challenges of Spot Quotation Buying:

- Limiting the number of invited suppliers may reduce competition, potentially impacting the aptitude to secure the best terms.
- The pre-qualification process may exclude potential suppliers who could have provided competitive offers.
- In a limited enquiry, the organization may miss out on innovative solutions or offerings from suppliers not included in the pre-qualification process.
- Suppliers who are not pre-qualified may perceive the process as unfair or biased, leading to dissatisfaction.

5.4.3 Tenders

A tender, also known as a Quotation, is a formal document issued by an organization seeking goods, services, or works from potential suppliers or contractors. The tendering process is a competitive procurement method used by public and private sector organizations to ensure transparency, fairness, and efficiency in selecting suppliers or contractors for various projects

Types of Tenders

- Single Tender
- Open Tender
- Limited Tender

5.4.3.1. Single Tender

Tender is invited from one reliable supplier only. Single Tender is called under following conditions.

- Priority items
- High Quality items
- C Class Items such as Clips, Pins, Pencils etc.
- When items are required comparatively urgent
- Bidders submit their complete bids in one stage.
- Common for straightforward projects.

5.4.3.2. Open Tender

Open tender which is also called press tender is published in Newspaper, trade Journals etc., For procuring materials of desirable specifications.

- Open to all qualified suppliers or contractors.
- Open tender gets widely publicity.
- Advertised publicly, and any concerned party can submit a bid
- A vendor has to deposit an earnest money with the tender information. This is just to confirm that the vendor does not back out from the rates.

5.4.3.3. Limited Tender

A limited tender is a procurement method where an organization invites a restricted or limited number of suppliers or contractors to submit bids for a specific project, contract, or procurement requirement. This approach falls between open tendering (where any interested supplier can participate) and closed tendering (where participation is highly restricted to a select group). Limited tendering is also known as selective tendering. Steps of limited tendering is shown in figure 5.9.

- Before issuing a limited tender, the organization typically conducts a pre-qualification process to identify and select potential suppliers.
- After pre-qualification, the organization invites a limited number of traders to participate in the tendering process.
- The invitation for a limited tender is project-specific, and suppliers are selected based on their suitability for the particular project or requirement.
- Details of the tender, including specifications and requirements, may be disclosed only to the invited suppliers to maintain confidentiality.
- Limited tendering allows for competition within a controlled group of suppliers, ensuring that there is a competitive element while maintaining a degree of exclusivity
- The organization is expected to conduct the limited tendering process transparently, ensuring that the criteria for trader selection are clear and communicated to all potential participants.

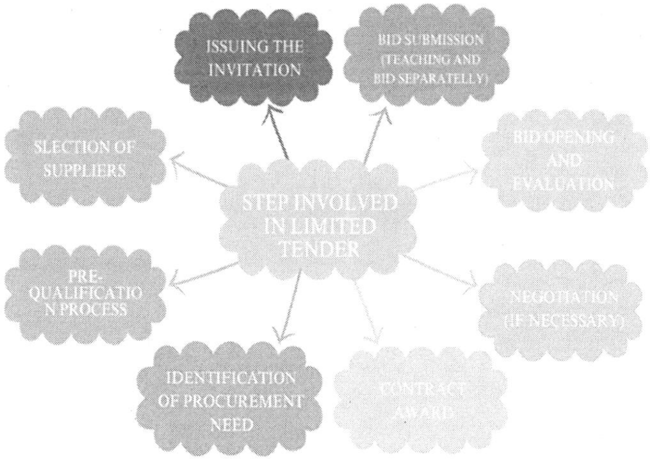

Figure 5.9: Steps involved in Limited Tender

A Sample format of tender is presented here for better understanding

Proposal for bidding of Limited Tenders

<u>Notice for tenders</u>

Limited tender No: Date:

Sealed limited tenders are requested from the Producers/ Suppliers/Authorized dealers/ agencies for the supply and installation of the following-

Table 1: List of Particulars with specifications

Sl. No	Particulars with Specifications	Quantity
1.		.... Nos

The tenderer shall be required to submit the Earnest Money Deposit (EMD) for antotalof Rs. .... by way of demand drafts/banker's cheque/FDRs which is refundable and a non-refundable tender fee for an amount of Rs. **200/- (Rupees two hundred only)** by Demand draft. The demand drafts (validity 45 days beyond final bid) for earnest moneycredit& tender fee must be enclosed in the envelope holding the bid documents, super-scribed with tender number, duedateofsubmissionon the envelop and addressed to :

The Registrar

..................................,

..................................,

.................."

Note: Central Purchase Association, Small Scale Industries/ National Small Scale Industries Corporation shall be exempted from expense of Earnest Money Deposit. Tenderer seeking exclusion should enclose a self-attested photocopy of valid registration certificate with NSIC.

(The Earnest Money will be liable to be forfeited if quotation is not honored or if contract is not signed with the Institute, after the grant is made to the Tenderer)

1. Time and end date of offer of the Bid: 11.00 am on 30.12.20...
2. Time of Bid Opening: 11.30 am on 30.12.20...
3. Venue of Bid Opening: ...., ... in the presence of bidders who want to be present at the time of opening of tender.

Interested bidders may post (at the above address) or put the tender papers completed in all respect and other requisite papers in the tender box kept in the General Section, CAO, The bidders are also informed that they may come personally or send their demonstrative to be present at the time of opening of bid. Please note that tender box shall be opened at the time mentioned above irrespective of whether bidders himself or any of their representative are present or not. The tenders shall not be entertained after this deadline under any conditions what so ever. For more details, kindly visit the Institute's website http://www............... or contact .... Ph. No....................

Registrar

General Terms & Conditions

Note: Bidders must submit the following primary information/documents with the quotation. Bidders will have to indicate these particulars in their quote failing which the offer may be rejected. Please do produce the related documents whenever required by the Institute.

1. Trade License/Company Registration No.
2. GST Regn. No.
3. Income Tax PAN No.
4. Firm's Bank A/c particulars
5. Bidders are requested to quote rate(s) per unit(s) only in the recognized Accounting units otherwise your quotation will not be accepted.
6. Cost of items shall include installation, support and troubleshooting, if any.
7. Warranty and Support: for Hardware and Software should be explicitly mentioned, if any.
8. Bidders should be OEM/Authorized partner/Authorized dealer of OEM for the item to be purchased in question.
9. Bidders should quote rates as per details/specifications mentioned in notice inviting Tender. The Institute reserves the right to place order for each item/job to single/separate vendor(s) if necessary. The quantity of the item(s) may vary.
10. Bidders must indicate if their rates are inclusive of Taxes.
11. In case opening date of Tender happens to be holiday, tender will be acknowledged and opened on the next working day at the same time and same place. Quotation received after the closing date will not be entertained and revision in the price will render the bid invalid. Quotation should indicate clearly the period of validity, preferably not less than 45 days.
12. In case of an offer for items having multiple options, vender should clearly indicate item-specific price(s). Please quote separate item-wise rate(s), when quotation has been asked for so. For every offer, packing and forwarding charges, GST or any other Tax etc. should be shown separately.
13. Bids will be evaluated after equated comparison of offers upon calculating all tax/duty/cess/surcharge/discount/packing/transportation costs, other charges with price etc.
14. and non-compliance of technical and commercial terms will render a bid liable for rejection.
15. Bidders will have to submit Bills/Invoices on dispatch of stores, if ordered, to this office in triplicate duly pre-receipted (and stamped for amount over Rs. 5000/-) and supported by the relevant delivery documents for audit and payment directly in your bank account through RTGS/NEFT. Generally, payments can be expected within one month and are made against acceptance of supplies/ jobs completed and in deserving cases, against shipment credentials.

Registrar

Method to short list a best suitable vendor

Purchase above Rs. 1.00 lac and upto Rs. 25.00 lac.

Minutes of the Purchase/Equipment Committee meeting held on_08.01.20…._ in ………..at 12 noon

The following members were present:

1. ………
2. ………

The following item(s) were considered:

Details of Machine with specification

These items are to be used to setup a new Machine workshop at tannery to manage Turner, students, Personal Contact Programme workshops and increased strength of students at Institute.

The fund for the above item(s) is available from **……… *account grant*** and Financial / G.B. clearance (*Letter ………… dated …………..*) for above items has been obtained vide resolution 41(f) dated ………… Limited tender with the approval of Director were invited. Notice inviting quotations was put up on official website on …………… Tender was opened on ……….. at 11.30 am in Room No.3.

Following quotations were received:

Vender 1 :

Vender 2 :

Vender 3 :

S.no.	**Details of Item with Specification**	**The Comparative statement of quotations**				
		Vendor 1	**Vendor 2**	**Vendor 3**	**Vendor 4**	**Vendor 5**
1	Basic Price per unit					
2	GST					
3	GST Charges					
3	Transit Insurance Charges					
3	Installation charges					
4	Transportation charges					
5	Packaging and forwarding charges per machine					
6	**Total Price for Machines including GST and all other Expenses**					

7	Payment terms Condition					
8	Warranty/guarantee					
9	Vailidity					
10	**Remarks and Justification**	***M/s .... (Name of Vendor) quoted lowest quotations among 5 bidders.***				

Table 5.1: Comparative Statement of Quotations

Resolved that the following quotations be accepted for purchase:

Resolved further that ……….. is hereby authorized to place order for the supply of following:

Item	**Vendor**	**Net Price (INR) including all taxes**	**Total Qty**	**Total amount (INR approx.)**
			Grand Total	

Table 5.2 Detail of vendor for Purchase order

Also recommended that advance be drawn in the name of ………….. for wire transfer.

(Director or nominee) (Treasurer or nominee) (Head of Department) (Dean/Principal/Registrar)

5.4.4 Terminology in tenders

5.4.4.1. Earnest Money

Earnest money, also known as earnest deposit or bid security, is a financial commitment made by a bidder or a party to demonstrate their serious intent and commitment when participating in a tender process or entering into a contract. The aim of earnest money is to provide assurance to the party accepting bids (usually the buyer or project owner) that the bidder is genuine and will fulfill their obligations if their bid is accepted.

- Earnest money is a symbol of good faith and demonstrates the bidder's genuine interest in the tender process or contract.
- In the context of tenders, earnest money serves as bid security.
- The terms regarding the refund ability of earnest money vary based on the terms of the tender or contract.
- It is often a small percentage of the bid amount, but the specific requirements vary by industry and region.

5.4.4.2. Security Deposit

After selecting the supplier to whom give the tender, either on the basis of lowest rates quoted by him or otherwise, he is asked to make a security deposit so that in case the supplier fails to furnish the goods properly and in time, the security deposit can be forfeited.

5.5 Ordering of Material or the Purchasing Procedure

Ordering involves a series of steps and activities (Refer Figure 5.10) aimed at acquiring goods or services needed to meet the organization's operational requirements. .

- Departments or units within the organization identify their material requirements based on production needs, project demands, or any other operational considerations.
- A formal document known as a purchase requisition is generated to specify the materials or services needed, along with applicable details such as quantity, specifications, and delivery requirements
- The purchase requisition undergoes an authorization and approval process. This involves getting sanction from relevant authorities within the organization, such as department heads or budget managers.
- The procurement team or purchasing department identifies possible suppliers who can provide the required materials or services.
- Depending on the nature of the purchase, an request for quotation or request for proposal may be issued to selected suppliers. This document outlines the organization's requirements and solicits pricing, terms, and conditions from suppliers
- The procurement team evaluates the received quotations or proposals based on various factors, as well as price, quality, delivery time, and any other relevant criteria.
- Based on the evaluation, a supplier is selected. This decision may involve negotiations with the chosen supplier to finalize terms and conditions.
- Once the supplier is selected, a purchase order is created. The PO is a formal document that details the agreed-upon terms, including the quantity, specifications, price, delivery date, and payment terms.
- The supplier confirms the acceptance of the purchase order and acknowledges the terms and conditions specified in the document.
- Upon receiving the materials or services, the receiving department checks the delivery against the information in the purchase order. A goods receipt is created to confirm that the items were received in good condition and in the specified quantities.
- The finance or accounts payable department verifies the supplier's invoice against the purchase order and goods receipt to ensure accuracy and compliance with the agreed-upon terms.
- Once the invoice is verified, the organization processes the payment to the supplier according to the agreed-upon payment terms.

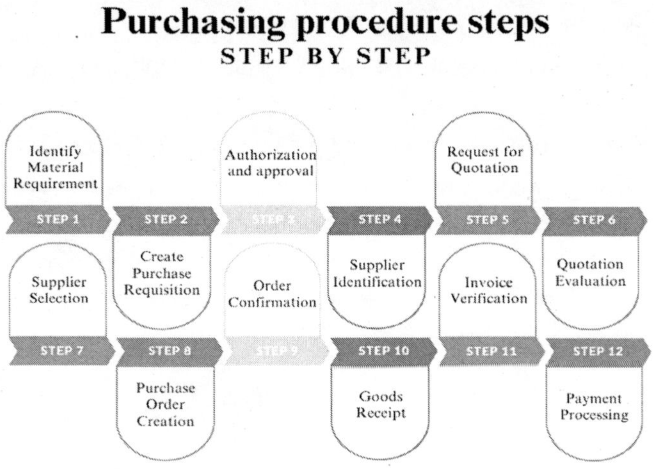

Figure 5.10: Steps of Purchasing Procedure

5.6. Stores and Storekeeping

5.6.1 Introduction

Store and storekeeping are a service to production department. Stores and storekeeping are integral components of the supply chain and inventory management within an organization. The terms "stores" and "storekeeping" refer to the physical locations and activities associated with the storage, control, and distribution of goods or materials.

In a business context, store typically refers to the physical facilities or locations where goods or materials are kept, organized, and managed. Storekeeping refers to the set of activities, processes, and practices involved in the management and control of goods within a store.

In the word of Maynard, the duties of storekeeper areto receive materials, to protect them while in storage from damage and unauthorized removal, to issue the material in right quantities, at the right time to the right place and to provide these services promptly and at least cost".

5.6.2 Categories of stores

Stores are generally of the following categories.

- Raw Material and Consumable material Stores
- Inflammable Material Stores like petrol, diesel, Oil etc.
- Tools Store
- Chemical Store Like sulphuric acid, caustic soda and carbide etc.
- Stationary Store
- Store to keep furniture
- Foundry item Store
- Electrical Goods store
- Scrap Store
- Finished Product store

5.6.3 Functions of storekeeping

- Receiving incoming shipments, inspecting goods for quality and quantity, and documenting the receipt.
- Picking and packing items to fulfill orders or requisitions from different departments.
- Managing inventory to ensure that older stock is used or sold first to prevent obsolescence.
- Implementing security measures to protect stored goods from theft or damage.
- Maintaining accurate records of inventory levels, transactions, and other relevant information.

Figure 5.11: Objective of Store keeping

5.6.4 Duties of Storekeeper

Most important duty of a storekeeper is the systematic planning of stores.

- Verifying and documenting incoming shipments, checking for damages, and ensuring that the received goods match the specifications in purchase orders.
- Efficiently organizing goods within the store to optimize space and facilitate easy retrieval. This involves implementing appropriate shelving, labeling, and tracking systems.
- Picking, packing, and dispatching goods based on orders or requisitions from different departments or units.
- Regularly monitoring and updating inventory levels, conducting physical stock counts, and implementing measures to prevent stockouts or overstocking.
- Ensuring that stored goods meet quality standards and taking appropriate actions in case of damaged or defective items.
- Implementing security procedures to protect against theft, unauthorized access, or damage to stored goods.
- Maintaining accurate and up-to-date records of all transactions, including receipts, issues, and stock balances.

5.6.5 Store Functions

In the industry, major store functions typically revolve around managing the flow of materials and goods within the production process. These functions are critical for ensuring efficient operations and meeting customer demand. Here are some of the key store functions in manufacturing:

Figure 5.12: Store Functions

5.6.6 Store Organization

Generally, the following two kinds of organization are approved in relation to stores

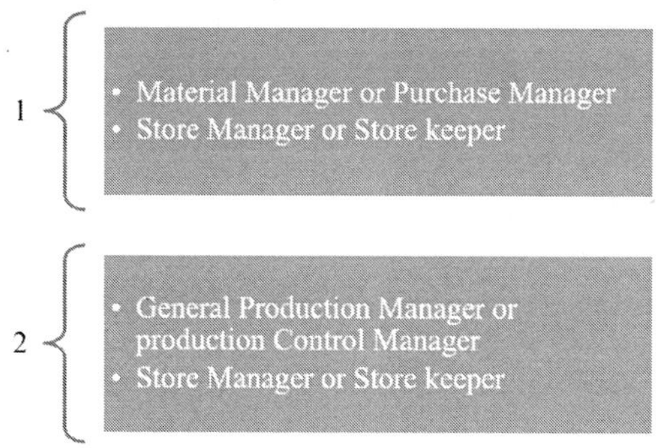

Figure 5.13 Store Organization Structure

Stores, in the context of inventory management and storage facilities, can be organized in different ways based on the distribution of control and decision-making. The two main types are centralized stores and decentralized stores.

5.6.7 Centralized Stores

In a centralized stores system, there is a single, central storage facility that manages and controls the inventory for the entire organization.

5.6.7.1 Advantages of Centralized Store

- ➢ Centralized storage can lead to economies of scale, resulting in potential cost savings.

- Easier implementation of standardized processes and procedures across the organization.
- Centralized stores allow for the concentration of expertise in inventory management.
- Centralized stores can lead to more efficient inventory management, as there is better visibility and coordination.

5.6.8 Decentralized Stores:

In a decentralized stores system, multiple storage facilities are distributed throughout the organization, and each department or production unit manages its own inventory.

5.6.8.1 Advantages of Decentralized Store

- Materials are available for departments, leading to shorter lead times.
- Each department can adapt inventory management practices to its specific needs.
- Reduced transportation costs for materials moving between departments.
- Each department or unit has regulator over its own inventory management, ordering, and distribution.

5.6.9 Stores Systems and procedures

The following headings provide a general framework for studying store systems and procedures:

5.6.9.1 Identification of Materials

When large number of dissimilar items are stocked in the store room, it is essential to codify them for their quick identification. "There should be a place for everything and everything should be in its place". Provision of a good identification system in stores assists those who try to keep everything in place. Identification can be done by: -

- Tagging some piece of paper or cloth with the items.
- Fixing labels on the items.
- Painting color coding of items.
- The coded number or any other identification mark may be embossed on the items.

Coding is a technique of assigning brief name to the items stocked in the store room. Generally, alphabets, numerals or a combination of the two is used for codification of items. The following procedures can be used for code construction:

I. **Mnemonic Method**: In this method alphabets closely associated with the name of the item are used e.g. MT for metallic items, PT for plastic items etc. This method is useful when few types of items are to be stored.

II. **Random Method.** Under this method, the numerical and alphabets are used on random basis. It is flexible as it is possible to add the new number or letter in sequence in case of addition of new items. But this method is rather arbitrary.

III. **Scientific Method.** Under this method, the total stock of items is divided into number of groups or classes and a symbol is assigned to each group. Each group may be further sub-divided until the individual item is identified.

Location Coding. Store room is divided into blocks of storage units and each block is identified by lateral block letter and longitudinal block letter. Within each block every row of shelves is given a number. Each row is divided vertically into columns and horizontally into shelves. Blocks and Rows are identified by painted signs.

Location coding is desirable especially where the huge number of items are stocked in big godowns located at different places. The same principles of code identification are followed while constructing the location code. The location is identified in terms of number of warehouse (if more godowns are maintained), row number, column number, rack number, shelf or bin number etc.

5.6.9.2 Receipt of material

Materials, tools, equipment are shipped by the supplier along with the necessary documents and packing slip. The storekeeper unloads the goods and verifies the contents with the packing slip and the purchase order. Before storing, these items are subjected to inspection or test to ensure that they are affording to the required specifications as to both quantity and quality as stated in the purchase order. A goods acknowledged note or good inward sheet is then prepared for these articles. A specimen form of goods inward sheet is given below:

GOODS INWARDS SHEET

ABC CO. LTD. MUMBAI

Sheet No. **Order No.**

Date Received

Invoice No.	Name of Supplier	Method of Delivery	item No.	No.	Weight	Dimensions (Spec.)	rate	Value	Condition	Qty. in order	Qty. Received	Remarks

Inspected and passed……. Received

Signature …………………… Signature

…………………..

Chief Storekeeper ………….. Receiving

Inspector ………………….. Clerk

Table 5.3 Receipt of Material

5.6.9.3 Storage of Materials

The most important function of the store department is to store and preserve the materials, tools, supplies etc. till they are issued to the production and other departments. These materials should be stored in suitable containers. Dissimilar types of containers such as drums, pallets, boxes, flexible racks, shelves, bins etc. are used for storing different types of materials. Some materials are also stored on floor in the form of heaps, e.g., molding sand, coal, iron ore etc.

Principles of Storing. Efficient control of stores is possible by applying following principles:

All the materials must be recognized, labeled and coded appropriately. When material codification is convoyed by location codes, it facilitates exact location of the materials inside the storeroom.

(i) All the bins, racks, shelves etc. must have signs and tags attached to them. Materials should be stored at proper safe place to prevent these from damage or mishaps.

(ii) Each item stored must bear name, class and reference number. It helps in easy and quick identification. It rejects chances of duplication of items likely to be presented under different names. It also helps in standardization and reducing the varieties of the materials.

(iii) Floor plan must be provided at convenient places in store room. It eliminates waste and helps in easy location of material.

(iv) The storeroom should be divided into sections and supplies belonging to particular class should be stored in it.

5.6.9.4 Issue of Materials.

Materials should be issued by the storekeeper only on the presentation of an indent, known as Stores Issue Requisition. The requisition is usually signed by the foreman but in few cases where extra large quantities are required for production, approval of higher authority may be required

<table>
<tr><td colspan="7">MATERIAL ISSUE REQUISTION (ORINDENT)</td></tr>
<tr><td colspan="7">Material Required
For Shop/dept...........Order No............... Indent No ......Date..........</td></tr>
<tr><td rowspan="2">Sr. No.</td><td rowspan="2">Materials with Specification</td><td colspan="2">Quantity</td><td rowspan="2">Rate</td><td rowspan="2">Enter Store Register No.</td><td rowspan="2">Remarks</td></tr>
<tr><td>Demanded</td><td>Supplied</td></tr>
<tr><td></td><td></td><td></td><td></td><td></td><td></td><td></td></tr>
<tr><td colspan="7">Requisitioned By .... Approved By........... Material Issued By .... Received By ....</td></tr>
</table>

Table 5.4 Material Issue Requisition

Such indents on stores are made out in triplicate form bound books and are supplied to each department. The original copy is sent to cost department for costing, duplicate retained by the storekeeper and triplicate in the bound book by the indenter by way of permanent record.

5.6.9.5 Record Keeping

In order to maintain an efficient record of stores, the following books and records will be required and these are maintained distinctly for different types of materials:

(i) Bin card.
(ii) Stores record card.
(iii) Material transfer note.
(iv) Material return note.
(v) Inward and Outward Register.
(vi) Stock registers
- Dead stock register
- Consumable register
- Tools Register

BIN CARD

Bin No.				**Min. Quantity**
Article				**Ordering level**
Code No...............				
Date	**Qty. Received**	**Qty.issued**	**Balanced**	**remarks**
Checked by............				Date...........

Table 5.5 Stores Accounting and Verification Systems

5.6.9.6 Methods of Pricing the materials issued or valuation of materials issued from stores

For the purpose of costing the receipt of materials, the factors that should be included are material price freight charges, insurance, duties, taxes and packing charge etc. The price quoted and accepted in purchase order may often be stated in various ways such as net prices, prices with discount term, free on board cost, insurance, freight etc. All these factors should be appropriately accounted for while costing for the incoming materials.

Another important accounting is to be done for the issue to production and of the stocks held at the end of accounting period. Let us discuss some of the important and frequently used system for this purpose:

(a) FIFO System. The 'First in First Out' approach operates under the premise that the oldest supply runs out first. Consequently, the applicable rate will be imposed at the time of issuing. The price arrangements don't involve any "profit" or "loss." The amount of money paid for that quantity of stock at the most recent price levels is the worth of the stocks that are kept on hand. The FIFO System gets cumbersome when price levels fluctuate too frequently. This system's inability to satisfactorily address costing-returns from retailers is another drawback.

(b) LIFO System. The 'Last in First Out' approach operates under the premise that the receipts that are issued the most recently are the ones that are used. This technique charges the most recent pricing, which results in reduced reported profits during price increases and potential tax savings. This technique has a tendency to shield unrealized gains or inventory losses from large price swings. Its constraints are nearly identical to those of the FIFO System.

(c) Average Cost System. This is dependent on the idea that the manufacturing department receives equal contributions from various shipments that are in stock; that is, a charge for the average cost of shipping to the stores. It keeps the cost figures steady. The average should be updated with each new purchase and is computed by dividing the total cost by the quantity of products.

(d) Market Value System. This method, where the materials are given at the prevalent market rates, is sometimes referred to as replacement rate costing. When prices rise, this method undervalues the amount of stock on hand; when prices fall, it overestimates the amount of stock on hand. This could therefore result in writing off a significant sum to make it more reasonable. Furthermore, the approach is laborious because it requires constant monitoring of the market prices for every resource.

(e) Standard Cost System. With this technique, a standard rate is established for a predetermined time frame, such as around six months, after a thorough review of market price and trends. Regardless of the actual rate, this standard rate is applied to materials issued during this time. Following the expiration of the time, the standard rate is examined and revised.

The effective utilization of resources is reflected in this system because accounting does not take rate fluctuations into consideration. Additionally, since the fresh rates don't need to be collected every time, it increases clerical efficiency. But in the event of rising or decreasing prices, respectively, this also results in underestimating or overestimating the stocks on hand, just like the Market Value Approach does.

(f) System of Costing the Closing Stock. Market price or stock at cost, whichever is lower, should be used as a general rule of thumb. Obsolescence, degradation, and price units are the key factors influencing closing stock costs. The stock may, under exceptional circumstances, increase in value over time. It is important to consider previous experiences while developing appropriate formulas to account for these aspects.

5.6.10 Stock Verification Systems

Some discrepancies between the actual and the book balances of inventories are bound to occur despite the diligent storekeeping. The process of stock verification is carried out for following purposes:

(i) To reconcile the store records and documents for their accuracy and usefulness.

(ii) Identification of areas deserving tighter document control.

(iii) To back-up the balance sheet stock figures, and

(iv) To minimize the pilferage and fraudulent practices.

To account for these disparities, the majority of businesses maintain a "inventory short and over" account, which is eventually closed into the manufacturing overheads account. The following are a few physical stock taking systems:

Annual or Periodic Physical Verification: In this approach, at the conclusion of a period—typically the accounting period—the entire inventory is physically inspected. In other words, usually at the end of the fiscal year. For several days, stocks are closed. The actions like equipment and machinery repair and overhaul are resorted to; this can require stopping industrial operations. Every item is physically inspected by a dedicated team of store inspectors and stores verifying officers, who are typically from the material audit, who also compare the entries on the bin card with the store ledger. A list of excess or short items is the result of this. Items that are broken or outdated are located and documented. To finish the verifications, both store- and item-wise, a thorough schedule and programming must be developed. Then, in order to write off shortcomings or value surplus, top management approval might be requested. The following items are all examined at once, so there is no need to worry about any being overlooked.

Perpetual Inventory and Continuous Stock Taking System: The final inventory system may take a long time for large businesses handling a lot of things, and it might not be able to shut down the entire factory. For larger factories, the permanent inventory system is a better fit. This approach involves ongoing stock checking all year round. Various companies use different approaches for ongoing verification.

Certain companies split their entire inventory into fifty-two equal portions. Every week, each component is validated. A number of products are counted daily or at regular intervals and cross-checked with the bin cards and stores ledger. Some businesses record store balances following every receipt and issue. Any discrepancies that are discovered—whether they are the result of broken items, theft, over-issuing, erroneous item placement in the wrong bin, etc.—are looked into and fixed appropriately. The following are this system's main benefits:

- It is not required to shut down the plant in order to verify or take stock
- This makes the process more accurate because it is less expensive, time-consuming, and tiresome.
- Store anomalies and flaws are easily identified and are not persistent throughout the year.
- This keeps things from getting damaged or lost.

- Slow-moving inventory can be identified so that appropriate action can be taken quickly.
- Stock items are kept within the allotted amounts.

Low Point Inventory System: Some corporations take the physical inventory, i.e. the stock level of stores is checked generally when it reaches its minimum level.

5.6.11 Material Storage Equipment and Filing System

I. **Shelves.** Shelves are used for storing of items to be issued in small lots. Shelves may be of wood or steel.

II. **Bins.** These like shelves are used for storing materials which are issued in small lots and cannot be economically palletized.

III. **Racks.** Made up of wood, steel, plastic or rubber, racks are used for the storage of materials of large dimensions.

IV. **Pallets.** Made up of wood, pallets are elevated platforms with lower under clearance and are used for stacking drums, lumber, rods, pipes and castings.

V. **Cardex Filing and Suspended Filing.** In the cardex filing. small sized cards (12.5 cms x 7.5 cms) are used to record all the relevant information about various materials in the stores and these cards are classified (as per the characteristics of the article) and arranged alphabetically for easy and quick reference. Suspended filing is an example where a lage volume of filing can be located in the same area and at the same time makes it easier to withdraw files.

5.7 Inventory

Inventory is a comprehensive record of the movable items required for product manufacturing and equipment maintenance. Every item in the list includes both its quantity and value.

Inventory is actually 'money' kept in the store room in the shape of a high-speed steel bit, a mild steel rod, milling cutters or welding electrodes

5.7.1 Inventory Control

Inventory control is concerned with achieving an optimum balance between two competing objectives. The objectives are:

I. To minimize investment in inventory,

II. To maximize the service levels to the firm's customers and its own operating departments.

Inventory control is the systematic approach to determining the optimal stock levels needed to meet production demands and ensure the timely availability of materials in the right quantities and at competitive prices.

5.7.2 Inventory Classification

Inventory may be classified as follows:

I. **Raw inventories**: They include, raw material and semifinished products supplied by another firm and which are raw items for the present industry.

II. **In-process inventories:** They are semi-finished goods at various stages of manufacturing cycle.

III. **Finished inventories**: They are the finished goods lying in stock rooms and waiting dispatch.

IV. **Indirect inventories**: They include lubricants and other items (like spare parts) needed for proper operation, repair and maintenance during manufacturing cycle.

5.7.3 Inventory Management

- To **manage** these various kinds of inventories two alternative control procedures can be used

5.7.3.1 Order point system

- This has been the traditional approach to inventory control. In this system, the items are restocked when the inventory levels become low.
- Lot size and reorder point calculations are the more spectacular aspect of inventory management. Once the calculations are complete, the routing commences for checking deliveries and physical count the amount on hand.

5.7.3.2 Materials Requirements Planning (MRP)

- MRP is sometimes thought of as an inventory control procedure. It is really more than that.
- MRP is the technique used to plan and control manufacturing inventories.
- MRP is a computational technique that converts the master schedule for end products into a detailed schedule for the raw material and components used in the end products.

The detailed schedule identifies the quantities of each raw material and component item. It also tells when each item must be ordered and delivered so as to meet the master schedule for the final products.

It is important that the proper control procedure be applied to each of the four types of inventory as explained earlier.

In general, MRP is appropriate control procedure for inventory type (i) and (ii) (i.e. raw materials, purchased components and in-process inventory).

Order point systems are often considered as the appropriate procedure to control inventory types (iii) and (iv) (i.e. finished goods, maintenance and repair parts, cutting tools and fixtures, plumbing supplies etc.).

5.7.4 Inventory control, its objective and how to achieve them

Inventory control aims at keeping track of inventories. In other words, inventories of required quality and in desired quantities should be made available to different departments as and when they need. This is achieved by,

I. Purchasing material at an economical price, at proper time and in sufficient quantities so as not to run short of them at any instant.

II. Providing a suitable and secure storage location.

III. Providing enough storage space.

IV. A definite inventory identification system.

V. Adequate and responsible store room staff.
VI. Suitable requisition procedure.
VII. Up-to-date and accurate record keeping.
VIII. Periodic inventory checkup.
IX. Division of inventory under A, B and C items, exercising the control accordingly and removing obsolete inventory.

A good control over the inventories offers the following Advantages

I. One does not face shortage of materials.
II. Materials of good quality and procured in time minimizes defects in finished goods.
III. Delays in production schedules are avoided.
IV. Production targets are achieved.
V. Accurate delivery dates can be ascertained and the industry builds up reputation and better relations with customers.

5.7.5 Functions of Inventories

Inventory serves several functions within a business, particularly in the context of manufacturing and retail operations. Here are some key functions of inventories:

I. Separate different operations from one another and make them independent, so that each operation (starting from raw material to finished product) can be performed economically. For example, ordering of raw material can be carried out independently of the finished goods distribution and both of these operations can be made low cost operations say by ordering raw material and distributing finished goods in one big lot, than in small batch sizes. Besides economy, the men and machinery also can be better utilized if the operations are separated and carried out in various departments than if coupled and tied at one place.
II. Maintain smooth and efficient production flow.
III. Purchase in desired quantities and thus nullify the effects of changes in prices or supply.
IV. Keep a process continually operating.
V. Create motivational effect. A person may be tempted to purchase more if inventories are displayed in bulk.

5.7.6 Economic Order Quantity

A problem which always remains is that how much material may be ordered at a time. An industry making bolts will definitely like to know the length of steel bars to be purchased at any one time. This length of steel bars is called Economic Order Quantity any one time permits lowest cost per unit and advantageous. Before calculating economic order quantity, it is necessary to become familiar with terms like maximum inventory, minimum inventory, standard order and reorder point, which are known as Quantity Standards. **Figure 5.14** shows different quantity standards.

Starting from an instant when inventory OA is in the stores, it (inventory) consumes gradually in quantity from A along AD at a uniform rate. It is pre known that it takes L number of days between initiating order and receiving the required inventory. Therefore, as the quantity reaches point B, purchase requisition is initiated which takes from B to C, that is time R. From C to D is the inventory procurement time P. At the point D when only reserve stock is left, the ordered material is supposed to reach and again the total quantity shoots to its maximum value, i.e. the point A'(A=A').

Maximum Quantity OA is the upper or maximum limit to which the inventory can be kept in the stores at any time. Minimum Quantity OE is the lower or minimum limit of the inventory which must be kept in the stores at any time. The purpose should be to hold enough and not excessive stock of material. Stock holding:

a. Avoids running out of stock.
b. Helps creating a buffer stock which may be utilized if the material falls below the minimum level.
c. Makes sure the pre decided delivery dates.
d. Provides quick availability of materials.
e. Takes care of price fluctuations and shortage of inventory in the market.
f. Advises regarding, obsolete and slow-moving items.
g. Helps in standardization and thus reducing the variety of items to be handled.

Standard Order. (A'D) is the difference between maximum and minimum quantity and it is known as economical purchase inventory size.

Reorder Point (B) indicates that it is high time to initiate a purchase order and if not done so the inventory may exhaust, and even reserve stock utilized before the new material arrives. From B[1] to Do it is as lead time (L) and it may be calculated on the basis of past experience. It includes:

a) Time to prepare purchase requisition and placing the order;
b) Time taken to deliver purchase order to the seller,
c) Time for seller (vendor) to get or prepare inventory; and
d) Time for the inventory to be dispatched from the ventor's end and to reach the customer.
e) Time, (a) above is known as requisition time (R) and (b) +(c) + (d) is the procurement time (P)

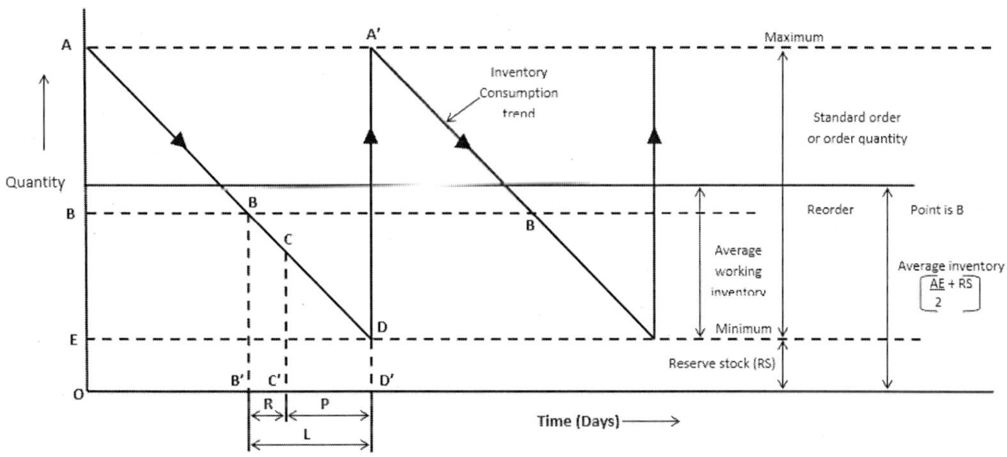

Figure 5.14: Economic Order Quantity

The economic lot size for an order or the economic order quantity depends upon two types of costs:

a) ***Inventory procurement costs**, which consist of expenditure connected with*

1. Receiving quotations;
2. Processing purchase requisition;
3. Following up and expediting purchase order;
4. Receiving material and then inspecting it; and
5. Processing seller's (vendor's) invoice.

Procurement costs decrease as the order quantity increases (see Fig.5.15)

b) ***Carrying costs**, which vary with quantity ordered, based on average inventory and consist of:*

Interest on capital investment;

1. Cost of storage facility, up-keep of material, record keeping etc;
2. Cost involving deterioration and obsolescence; and
3. Cost of insurance, property tax, etc.

Carrying costs are almost directly proportional to the order size or lot size or order quantity,In Fig. 5.15 the procurement cost and inventory carrying cost have been plotted with respect to quantity in lot. Total cost is calculated by adding procurement cost and carrying cost. Total cost is minimum at the point A and thus A' represents the economic order quantity or economic lot size. In Fig. 5.15. (b)the procurement costs and inventory carrying costs have been plotted with respect to Another method of finding E.Q.Q. that is by mathematical means, is given below:

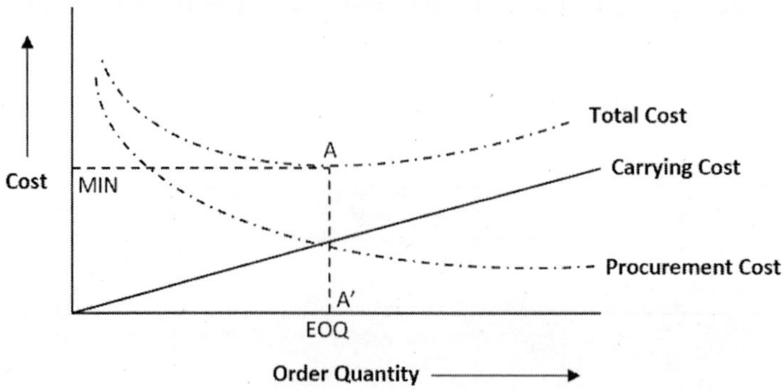

Figure 5.15: Relationship Between Cost and Quality

Let

Q is the economic lot size or E.O.Q.

*C*is the cost for one item.

*I*is the cost of carrying inventory in percentage per period, including insurance,

Obsolescence, taxes etc.

P is the procurement cost associated with one order.

and *U* is total quantity used per period say annually.

Number of purchase orders to be furnished

$$= \frac{TotalQuantity}{E.O.Q} = \frac{U}{Q}$$

Total procurement cost= Number of purchase orders x cost involved in one purchase orprocurement

$$= \frac{U}{Q} \times P$$

Average annual inventory $= Q/2$

Inventory carrying cost = Average inventory x cost per itcm x cost of carrying inventory in percent per Period.

$= \frac{Q}{2} \times C \times I$...(b)

Total cost, T = (a) + (b)

$$T = \frac{U \times P}{Q} + \frac{Q}{2} \times C \times I$$

$$T = U.P.Q^{-1} + \frac{Q}{2} \times C \times I$$

To minimize the total cost, differentiate T, w.r.t, Q and put it equal to zero

$$\frac{dT}{dQ} = \frac{d}{dQ}\left(U.P.Q^{-1} + \frac{Q}{2} \times C \times I\right)$$

$$0 = -U.P.Q^{-2} + C.I/2 \text{ or } U\,P/Q^2 = C.I/2$$

$$Q^2 = 2U.P.\ /C.I$$

$$Q = \sqrt{\frac{2U.P.}{C.I}} \qquad ...(i)$$

Example 5.1: Given that

i. Annual usage, U=60 units
ii. procurement cost, P=Rs.15 per order
iii. cost per piece, C=Rs. 100
iv. cost of carrying inventory I, a percentage including expenditure on obsolescence, taxes, insurance, deterioration etc.= 10%. Calculate *E.O.Q.*

Solution: $Q = \sqrt{\frac{2U.P.}{C.I}}$; substituting the values

$$Q = \sqrt{\frac{2\times6015}{100\times(10/100)}} = 13.41$$

Therefore, number of order per year= $\frac{60}{13.41} = 4.47$ say 5.

Hence Q or E.O.Q = $\frac{60}{5}$ 12 units (Ans.) (rounded)

The readers may try the following problems.

5.7.7 INVENTORY MODELS

- Inventory models determine when and how much inventory to carry.
- Inventory models handle chiefly two decisions.
    - How much to order at one time, and
    - When to order this quantity to minimize total costs.
- Lowest-cost decision rules for inventory management pertain to either buying products from outside or producing them within the company.
- Simple inventory models assume no delivery delay and that demand is known.
- Probabilistic models handle situations of risk and uncertainty.

5.7.7.1 Simple EOQ model

- ➢ The simple EOQ model can be used if the demand is known with certainty.
- ➢ The demand and lead time are known.
- ➢ The item will be purchased from outside (the firm) and that demand will continue well into the future.
- ➢ It is also assumed that not only the demand is known with certainty, but that is the same from day to day and that stockouts, are not allowed. Under these assumptions, Fig 5.16 depicts the inventory position through time.
- ➢ If we start to observe the inventory position immediately after receipt of an order, the quantity in stock *Q* decreases steadily until a lead time's supply is reached. If lead time (L) is 5 days and demand is 4 (pieces) per day, a lead time's supply is 20 pieces or units. Therefore when 20 units remain in the store, an order is placed. This is called the re-order pointer. Exactly 5 days after the order is placed, the stock is replenished and the cycle repeats itself.

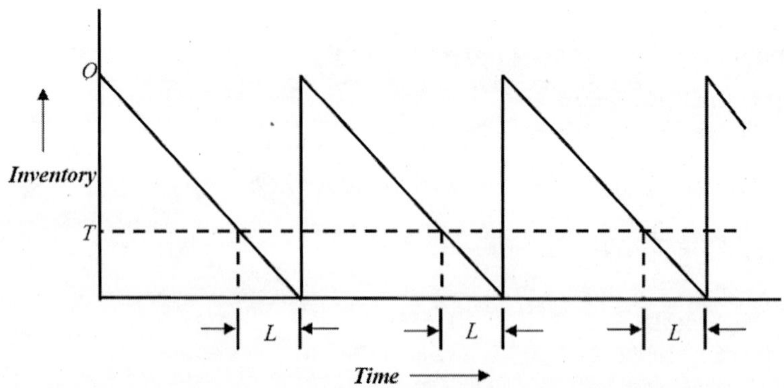

Figure 5.16 Time pattern for inventory level for simple EOQ Model.

5.7.8 A B C Analysis

5.7.8.1 Necessity

As the size of the industry increases, the number of items to be purchased and then to be taken care of their usage value, price or procurement problems, blocks and involves.a lot of money and man hours, of also increases. Purchase and control of all items at a time and in bulk much before their use, irrespective and is therefore uneconomical.

ABC analysis helps segregating the items from one another and tells how much valued the item is and controlling it to what extent is in the interest of the organization

5.7.8.2. Procedural Steps

1) Identify all the items used in an industry.
2) List all the items as per their value.
3) Count the number of high valued, medium valued and low valued items.
4) Find the percentage of high, medium and low valued items. High valued items normally contribute for 70% or so of the total inventory cost and medium and low valued items, 20 and 10% respectively.
5) A graph can be plotted between percent of items (on X-axis) and per cent of total inventory cost (on Y-axis). Figure 5.17 shows such a graph.

It can be seen that 70% of the total inventory cost is against 10% of the total items (called A-items), 20% against 20% of the items (B-items) and 10% against a big bulk, i.e. 70% of the items (called C-items). Thus, ABC analysis furnishes the following information:

1. A-items are high valued but are limited or few in number. They need careful and close inventory control. Minimum and maximum limits, and reorder point is set for A items. Such items should be thought of in advance and purchased well in time. A detailed record of their receipt and issues should be kept, and proper handling and storage facilities should be provided for them.

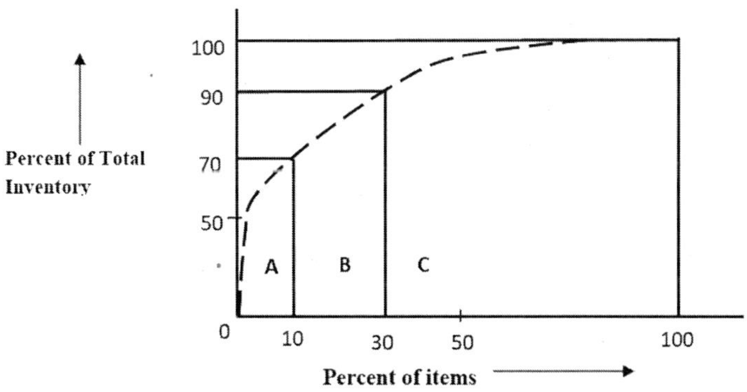

Figure 5.17: ABC Analysis

Such items being costly are purchased in small quantities oftenly and just before their use. This of-course increases the procurement costs and involves a little risk of non-availability. However, the locked Such items being costly are purchased in small quantities oftenly and just before their use. This of up inventory cost decreases and the problems of storage and care taking are minimized.

A-items generally account for 70-80% of the total inventory cost and they constitute about 10% ofthe total items.

2. ***B***-items are medium valued and their number lies in between A and C-items. Such items need moderate control. They are more important than C-items. They are purchased on the basis of past requirements; a record of receipts and issues is kept and a procurement order is placed as soon as the quantity touches reorder point. These items being comparatively less costly, a safety stock of up to 3 months may be kept, whereas it needs a stock of fortnight or so in case of A items. B-items also require careful storage and handling.

In brief, B-items need every care but not so intensive as is required for A-items.

B-items generally account for 20 to 15% of the total inventory cost and constitute about 15 to 20 % of the total items.

3. ***C***-items are low valued, but maximum numbered items.

These items do not need any control, rather controlling them is uneconomical. These are the least important items like clips, all pins, washers, rubber bands, etc. They are generally procured just before they finish. No expediting is necessary, no records are normally kept and a safety stock of 3 months or even more can be purchased at an instant. Future requirements of such items are never calculated and a two-bin system is sufficient to hint procurement.

C-items generally account for 10 to 5% of the total inventory cost and they constitute about 75% of the total items.

The factors affecting the importance of an item include annual usage, unit cost, and scarcity of material. For simplicity, only annual usage is used in this text. The procedure for classifying by annual usage is as follows:

The significance of an item is influenced by various factors such as annual usage, unit cost, and scarcity of material. However, in this particular text, only annual usage is taken into consideration. The classification procedure based on annual usage is outlined as follows:

- Determine the yearly consumption of each individual item.
- Compute the total annual consumption of each item by multiplying its annual usage with its cost.
- Arrange the items in order of their annual usage.
- Determine the total cumulative annual usage and the cumulative percentage of items.
- Analyze the distribution of annual usage and categorize the items into groups A, B, and C based on the percentage of annual usage.

Example 5.2: Lumax Pvt. Ltd. produces a range of 10 products. The table below displays the unit cost and usage of each item, as well as the annual usage calculated by multiplying the unit cost by the unit usage.

a) Determine the yearly consumption for each individual item.
b) Arrange the items in order based on their annual usage.
c) Compute the total annual consumption and the cumulative percentage of items.
d) Categorize the items into groups A, B, and C.

Answer

a. Calculate the annual usage for each item.

Part Number	Unit Usage	Unit Cost	Annual Usage
1	1100	2	2200
2	600	40	24000
3	100	4	400
4	1300	1	1300
5	100	60	6000
6	10	25	250
7	100	2	200
8	1500	2	3000
9	200	2	400
10	500	1	500
Total	5510		38250

b. b, c and d.

Part Number	Annual Usages	Cumulative Usages	Cumulative Usages	Cumulative % of items	Class
2	24000	24,000	62.75	10	A
5	6000	30,000	78.43	20	A
5	3000	33,000	86.27	30	B
1	2200	35,200	92.03	40	B
4	1300	36,500	95.42	50	B
10	500	37,000	96.73	60	C

Part Number	Annual Usages	Cumulative Usages	Cumulative Usages	Cumulative % of items	Class
9	400	37,400	97.78	70	C
3	400	37,800	98.82	80	C
6	250	38,050	99.48	90	C
7	200	38,250	100.00	100	C

Table 5.6 Categorization of items as per A, B, C classification

5.8 Vendor Development

Any manufacturing company will need a host of vendors for supply of raw materials and finished parts. The bigger the company and the larger the manufactured item, the more diverse the vendor profile. For example, a car manufacturing company would need vendors who are into metal components, rubber parts, upholstery, fasteners, oils. paints, tyres and so on. The car manufacturer would only manufacture a small number of critical items developed inhouse or assemble aggregates such as engine assembly out of parts supplied by various vendors.

Engaging vendors for supply of parts saves precious resources in terms of space, machinery, labour force etcetera. The company can concentrate on its core manufacturing and keep track of quality and targets without worrying about sundry inputs to achieve the same. Of course, it would need to engage and manage a large group of vendors skillfully to get good quality goods at competitive rates just in time for its assembly so as not to have to maintain a large inventory which would involve precious space.

Vendor development is an art and science in itself. Several points need to be kept in mind in cultivating a dependable group of vendors. These are enumerated below:

1. **Expertise**- Before finalizing a vendor, it is essential that its capabilities be assessed. This would entail a visit by the company's officials to the vendor's premises to see its plant and machinery to assess its manufacturing capability and quality control functions, worker and management profile, accounting procedures, strong financials etcetera.
2. **Locale**- A local vendor, that is one established in the same city or town as the company, would enable the company easy accessibility for inspection and quality control, competitive rates due to lower cost of transportation, better communication in times of urgency and so on.
3. **Cost of product**- There would be several vendors able to supply a particular item and the prices offered by each may vary due to inherent variables specific to each. It would be in the company's interest to select the vendor offering the item at the most competitive rate without compromising on quality and reliability of supply.

5.8.1 Vendor Selection

In order to select a dedicated team of dependable vendors, we have to go through a selection process. For this we float a Request for Proposal (RFP) and advertise it in local and national newspapers, through handbills, and by other means of advertising as per requirement and scale of our business so as to reach a wide audience and all suitable vendors who may be interested in doing business with us. We share with the responding vendors the specifications of goods and services required and call for quotations from each of them. These are then analyzed by an expert team of officials shortlist a probable list of vendors. We then have face to face discussions with the competitors to zero down on the final list.

Having analyzed the strengths and weaknesses of several vendors, we come down to selecting the desired number of vendors for an assortment of required functions.

5.8.2 Vendor Analysis

Vendor analysis is an important step in the vendor development process. You have to look for vendors who understand your company's needs and are geared to meet them in a timely and cost-effective manner.

The selected vendor must have the capability and the capacity to meet your requirement of goods and services. The important factors to consider include the strength and skill of its workforce, availability of suitable plant and machinery, uninterrupted power supply with adequate power backup, good management practices, a congenial overall work atmosphere, good worker-management relations, and strong financials. All these are vital to cater to any unforeseen variations in requirements of goods and services which would have to be responded with agility.

5.8.3 Vendor Rating

A vendor, aka supplier may be selected for supply of a particular item and its reassessment would have to be done after a certain period to evaluate its adherence to systems to maintain quality and performance. Hence, a system of vendor rating is introduced by which we assign a grade or ranking to various vendors. The rating is based on various parameters such as quality of goods, pricing, adherence to delivery schedules, and systems advancement by the vendor. The information shall be collected continuously through on-line portals so that data is not skewed and it is available in real time. Various parameters may be given different weightage, for example quality may be given the maximum weightage of 40% followed by delivery rating, say 25% and pricing and systems 20% and 15% respectively.

Then the overall vendor rating will be weighted average of the aforementioned parameters and can be determined by the formula given below:

VR=(40*QR+25*DR+20*PR+15*SR)/100

where

QR = Quality Rating

DR = Delivery Rating

PR = Pricing Rating

SR = System Rating

The quality rating (QR) of the product supplied is adjudged based on rejection during product inspection and its performance in the field. One always comes across news items mentioning recall of a good number of cars due to malfunctioning items. This not only costs money to rectify or replace the defective part but also earns a bad name for the manufacturing company. No one gets to know the name of the vendor which supplied the defective component.

The delivery rating (DR) of the product is arrived at by looking at the quantity delivered within the delivery period against the total ordered quantity and how much extra time is taken to deliver the complete consignment. Obviously, delivery of complete lot within the delivery period gives a DR of 100%.

The pricing rating (PR) is easier to comprehend because it is a straightforward comparison of rates quoted by various vendors and there is no subjectivity in this.

The systems rating (SR) is a reflection of the vendor's adherence to conditions of approval and company's directives agreed upon between the company and the supplier at the beginning as regards up gradation and maintenance of infrastructure and systems etcetera.

Example 5.3: M/S Denso and M/S Subros are both manufacturers of automobile passenger car air conditioning systems. Maruti Suzuki India Limited, an OEM, seeks to evaluate the performance of these vendors based on specified given parameters. Determine their respective ratings.

S. No.	Vendor	Rating out of 5			
		Quality Rating (QR)	**Delivery Rating (DR)**	**Pricing Rating (PR)**	**System Rating (SR)**
1	M/S Subros	4.6	4	4.2	4.5
2	M/S Denso	4.4	4.2	4.3	4

Table 5.7 Detail of vendor rating

Answer: To find out the ratings for M/S Denso and M/S Subros based on the given parameters, we can use the provided formula to calculate their Vendor Ratings (VR).

- VR=(40*QR+25*DR+20*PR+15*SR)/100
- For M/S Denso:
- VR_{Denso}=40×4.6+25×4.0+20×4.2+15×4.5100
- VR_{Denso}=4.355
- For M/S Subros:
- VR_{Subros}=40×4.4+25×4.2+20×4.3+15×4.
- VR_{Subros}=4.27

So, the Vendor Rating for M/S Denso is approximately 4.355 out of 5, and for M/S Subros, it's approximately 4.27 out of 5. These ratings provide an overall assessment of each vendor's performance based on the specified parameters.

5.8.4 Classification of Vendors

Based on pre-decided vendor rating (VR) slab rates, each vendor can be classified into different classes A, B, and C. In the case of vendors who maintain A rating throughout the period of approval, the validity can be extended for the next period of approval without the need for field visit by company officials. On the other hand, for vendors who are rated C class throughout the period of approval, the validity will not be extended. They may be issued a notice for improvement at first and if no improvement is discernible, a show cause notice for delisting can be issued. If there is no improvement in the subsequent year, the vendor can be finally delisted from Vendor List. A similar procedure can be decided for vendors with rating B. a hierarchy of vendors in a manufacturing industry is shown in figure 5.18

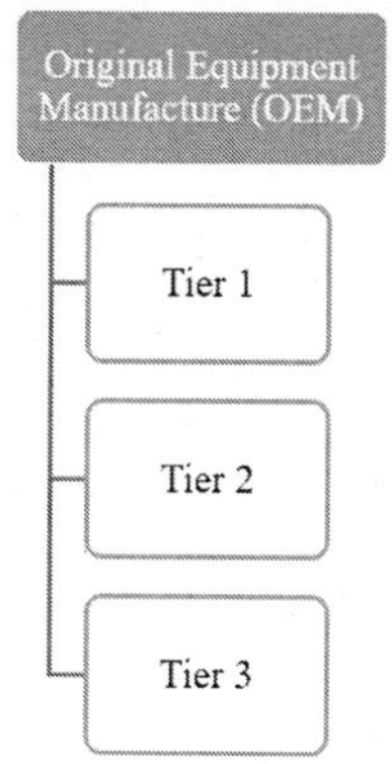

Figure 5.18 Hierarchy of Vendors in Manufacturing Industry

Vendor development, selection, analysis, and rating are important aspects in maintaining the quality and cost competitiveness of the final product. Since no company can operate in isolation due to so many constraints and increased liability if it tries to do so it has to depend upon a number of vendors to produce good quality parts at competitive cost so as to maintain its position in the market and grow with the times.

UNIT SUMMARY

This unit offers a thorough examination of material management, which encompasses a variety of critical components that are essential for engineering applications. The unit commences with the fundamentals of material management and then explores the functions of store and storekeeping, as well as purchasing and procurement. The discussion is comprehensive, providing a comprehensive understanding of the practical implementation and advantages of concepts such as tender processing, purchase process, and quotations analysis. Furthermore, the subject explores the industry's inventory control, vendor development, and economic order quantity. Additionally, a wide variety of exercises, such as numerical problems, multiple-choice questions, and short and long answer questions, are available to students, ensuring that they have a thorough understanding and mastery of the subject matter. This chapter enumerates each of these topics in detail.

EXERCISES NO: 5

Multiple Choice Questions

1. What is the primary goal of material management?
    a. Maximizing inventory levels

b. Minimizing production efficiency
c. Minimizing material costs
d. Maximizing lead time

2. Which of the following is a key component of material planning in supply chain management?
   a. Order fulfillment
   b. Inventory control
   c. Quality control
   d. Marketing strategy
3. What does ABC analysis in material management involve?
   a. Analyzing product cost structure
   b. Categorizing items based on their importance
   c. Assessing supplier performance
   d. Calculating economic order quantity
4. In material management, what is the Economic Order Quantity (EOQ)?
   a. The maximum order quantity for a product
   b. The minimum order quantity for a product
   c. The optimal order quantity to minimize total inventory costs
   d. The average order quantity for a product
5. Safety stock in material management is maintained to:
   a. Maximize inventory turnover
   b. Minimize carrying costs
   c. Protect against uncertainties in demand and supply
   d. Minimize order frequency
6. What is the purpose of a Bill of Materials (BOM) in material management?
   a. Tracking employee work hours
   b. Identifying the sequence of production activities
   c. Listing the components required to manufacture a product
   d. Calculating order quantities
7. Which inventory valuation method assumes that the latest items added to inventory are the first to be used or sold?
   a. FIFO (First-In-First-Out)
   b. LIFO (Last-In-First-Out)
   c. Weighted Average Cost
   d. Specific Identification
8. Just-in-Time (JIT) is a material management approach that emphasizes:
   a. Maximizing safety stock
   b. Minimizing lead times
   c. Maximizing order quantities
   d. Minimizing production efficiency

9. What is the primary purpose of a central store in an organization?
   a. Facilitating decentralized inventory control
   b. Minimizing transportation costs
   c. Concentrating inventory for better control and efficiency
   d. Reducing lead times
10. In a decentralized stores system, who typically has control over inventory management?
   a. Centralized authority
   b. Warehouse manager
   c. Local department or unit
   d. Procurement team
11. What is the key advantage of a decentralized stores system?
   a. Economies of scale
   b. Centralized decision-making
   c. Reduced lead times
   d. Standardization of processes
12. What is the purpose of a buffer store in a production environment?
   a. Reducing lead times
   b. Managing fluctuations in demand or production rates
   c. Concentrating finished goods for distribution
   d. Optimizing transportation routes
13. What is the primary goal of procurement in an organization?
   a. Maximizing costs
   b. Minimizing lead times
   c. Obtaining quality goods and services at the best value
   d. Reducing supplier relationships
14. What is the purpose of a Request for Proposal (RFP) in the procurement process?
   a. Placing a purchase order
   b. Inviting suppliers to bid on a project
   c. Receiving supplier invoices
   d. Managing inventory levels
15. In procurement, what does the term "Lead Time" refer to?
   a. The time it takes to negotiate with suppliers
   b. The time it takes to receive goods after placing an order
   c. The time it takes to process invoices
   d. The time it takes to conduct market research
16. What is the purpose of a Purchase Order (PO) in the procurement process?
   a. Requesting supplier quotes
   b. Authorizing payment to suppliers

c. Placing an order with a supplier
d. Negotiating with vendors

17. Which procurement document specifies the terms and conditions under which a supplier will provide goods or services to the buyer?
    a. Purchase Requisition
    b. Request for Proposal (RFP)
    c. Purchase Order (PO)
    d. Supplier Evaluation Form
18. In procurement, what is the purpose of conducting a make-or-buy analysis?
    a. Assessing supplier performance
    b. Evaluating whether to produce a component internally or purchase it from an external supplier
    c. Negotiating with vendors
    d. Conducting market research
19. What is the primary objective of inventory control?
    a. Maximizing order quantities
    b. Minimizing lead times
    c. Minimizing holding costs while ensuring product availability
    d. Maximizing safety stock levels
20. In inventory control, what is the purpose of the reorder point?
    a. Determining the optimal order quantity
    b. Identifying the point at which a new order should be placed
    c. Assessing supplier performance
    d. Estimating carrying costs

Answers of Multiple-Choice Questions

1 c, 2 b, 3 c, 4 c, 5 c, 6 c, 7 a, 8 b, 9 c, 10 c, 11 c, 12 b, 13 c, 14 b, 15 b, 16 c, 17 c, 18 b, 19 c, 20 b

Short and Long Answer Type Questions

1. Define Material Management and write down different Functions of Materials Management Department.
2. What is A-B-C analysis for inventory control? Describe in detail
3. Define the terms Inventory and Inventory control. Write down various important functions of inventory control.
4. Discuss the Inventory policies for A, B and C items.
5. What do you know about Inventory Carrying Cost? What factors contribute towards it?
6. Name various quantity standards used as a tool to control inventory.
7. What are essential steps in Inventory Control?

8. Write briefly about Procurement Cost.
9. What do you understand by Tier 2 vendor?
10. The rate of use of a particular raw material from stores is 20 units per year. The cost of placing and receiving an order is Rs. 40. The cost of each unit is Rs 100. The cost of carrying inventory in per cent per year is 0.16 and it depends upon the average stock. Determine the economic order quantity If the lead time is 3 months, calculate the reorder point.
11. A factory uses two pieces per day of a rod 6 mm in diameter and 150 mm long in one of their manufacturing processes. The rod costs Rs. 3 each and the total expenses involved in purchasing and receiving them are Rs. 50 per order. The annual inventory carrying cost per item is Re 1. The procurement period is 3 days and minimum stock kept is 8 pieces. Find out,
    i. Standard ordering quantity,
    ii. Reorder point, and
    iii. Maximum stock,

12. Find Economic Order Quantity from the following data:
    1. Average annual demand = 30,000 units
    2. Inventory carrying cost = 12% of the unit value per year
    3. Cost of placing on order = Rs. 70
    4. Cost of unit = Rs 2

KNOW MORE

REFERENCES

[1] *Kosturiak, J. and Gregor, M. (1995) Total production control. Production Planning and Control 6, 490-499.*

[2] *HounshelL, D. A., From the American System to Mass Production, 1800-1932, The Johns hopkins University Press, Baltimore, MD, 1984.*

[3] *Hyer, N. L., and U. Wemmerlov, Reorganizing the Factory: Competing through Cellular manufacturing, Productivity Press, Portland, OR, 2002.*

[4] *KAPP, K. M., & quot.; The USA Principle, & quot; APICS-The Performance Advantage, June 1997, pp. 62-66.*

[5] *VILLA, C., & quot.; Multi Product Assembly Line Balancing, & quot; Ph.D. Dissertation (Unpublished), University of Florida, 1970.*

[6] *Wang, Wei, et al. & quot.; Integration of Part Print Analysis in Industry 4.0 Manufacturing Systems. & quot.; Procedia Manufacturing, vol. 48, 2021, pp. 739-746.*

[7] *WHITNEY, D., Mechanical Assemblies, Oxford University Press, New York, 2004. [27] WILD, R., Mass Production Management, John Wiley & Sons, London, UK, 1972.*

-----------------*---------*

ANNEXURE

Appendix 1

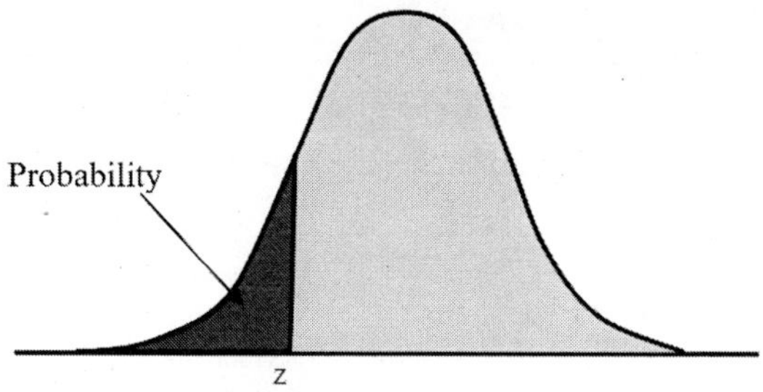

Table Entry for Z is the area under the Standard Normal Curve to the left of Z.

Standard Normal Probabilities

Z	.00	.01	.02	.03	.04	.05	.06	.07	.08	.09
-3.4	.0003	.0003	.0003	.0003	.0003	.0003	.0003	.0003	.0003	.0003
-3.3	.0005	.0005	.0005	.0004	.0004	.0004	.0004	.0004	.0004	.0004
-3.2	.0007	.0007	.0007	.0006	.0006	.0006	.0006	.0006	.0005	.0005
-3.1	.0010	.0010	.0010	.0009	.0008	.0008	.0008	.0008	.0007	.0007
-3.0	.0013	.0013	.0013	.0012	.0012	.0011	.0011	.0011	.0010	.0010
-2.9	.0019	.0018	.0018	.0017	.0016	.0016	.0015	.0015	.0014	.0014
-2.8	.0026	.0025	.0024	.0023	.0023	.0022	.0021	.0021	.0020	.0019
-2.7	.0035	.0034	.0033	.0032	.0031	.0030	.0029	.0028	.0027	.0026
-2.6	.0047	.0045	.0044	.0043	.0041	.0040	.0039	.0038	.0037	.0036
-2.5	.0062	.0060	.0059	.0057	.0055	.0054	.0052	.0051	.0049	.0048
-2.4	.0082	.0080	.0078	.0075	.0073	.0071	.0069	.0068	.0066	.0064
-2.3	.0107	.0104	.0102	.0099	.0096	.0094	.0091	.0089	.0087	.0084
-2.2	.0139	.0136	.0132	.0129	.0125	.0122	.0119	.0116	.0113	.0110
-2.1	.0179	.0174	.0170	.0166	.0162	.0158	.0154	.0150	.0146	.0143
-2.0	.0228	.0222	.0217	.0212	.0207	.0202	.0197	.0192	.0188	.0183
-1.9	.0287	.0281	.0274	.0268	.0262	.0256	.0250	.0244	.0239	.0233
-1.8	.0359	.0351	.0344	.0336	.0329	.0322	.0314	.0307	.0301	.0294
-1.7	.0446	.0436	.0427	.0418	.0409	.0401	.0392	.0384	.0375	.0366
-1.6	.0548	.0537	.0526	.0516	.0505	.0495	.0485	.0475	.0465	.0455
-1.5	.0668	.0655	.0643	.0630	.0618	.0606	.0594	.0582	.0571	.0559
-1.4	.0808	.0793	.0778	.0764	.0749	.0735	.0721	.0708	.0694	.0681
-1.3	.0968	.0951	.0934	.0918	.0901	.0885	.0869	.0853	.0838	.0823
-1.2	.1151	.1131	.1112	.1093	.1075	.1056	.1038	.1020	.1003	.0985
-1.1	.1357	.1335	.1314	.1292	.1271	.1251	.1230	.1210	.1190	.1170
-1.0	.1587	.1562	.1539	.1515	.1492	.1469	.1446	.1423	.1401	.1379
-0.9	.1841	.1841	.1788	.1762	.1736	.1711	.1685	.1660	.1635	.1611
-0.8	.2119	.2090	.2061	.2033	.2005	.1977	.1949	.1922	.1894	.1866

-0.7	.2420	.2389	.2358	.2327	.2296	.2266	.2236	.2206	.2177	.2148
-0.6	.2743	.2709	.2676	.2643	.2611	.2578	.2546	.2514	.2483	.2451
-0.5	.3085	.3050	.3015	.2981	.2946	.2912	.2877	.2843	.2810	.2776
-0.4	.3446	.3409	.3372	.3336	.3300	.3264	.3228	.3152	.3156	.3121
-0.3	.3821	.3783	.3745	.3707	.3669	.3632	.3594	.3557	.3520	.3483
-0.2	.4207	.4168	.4129	.4090	.4052	.4013	.3974	.3936	.3897	.3859
-0.1	.4602	.4562	.4522	.4483	.4443	.4404	.4364	.4325	.4286	.4247
00	.5000	.4960	.4920	.4880	.4840	.4801	.4761	.4721	.4681	.4641

Z	**.00**	**.01**	**.02**	**0.3**	**.04**	**.05**	**.06**	**.07**	**.08**	**.09**
00	.5000	.5040	.5080	.5120	.5160	.5199	.5239	.5279	.5319	.5359
0.1	.5398	.5438	.5478	.5517	.5557	.5596	.5636	.5675	.5714	.5753
0.2	.5793	.5832	.5871	.5910	.5948	.5987	.6026	.6064	.6103	.6141
0.3	.6179	.6217	.6255	.6293	.6371	.6368	.6406	.6443	.6480	.6517
0.4	.6554	.6591	.6628	.6664	.6600	.6636	.6672	.6808	.6844	.6879
0.5	.6915	.6950	.6985	.7019	.7054	.7088	.7123	.7157	.7190	.7224
0.6	.7257	.7291	.7324	.7357	.7389	.7422	.7454	.7486	.7817	.7549
0.7	.7580	.7611	.7642	.7673	.7704	.7734	.7764	.7794	.7823	.7852
0.8	.7881	.7910	.7939	.7966	.7995	.8023	.8051	.8078	.8106	.8133
0.9	.8159	.8186	.8212	.8238	.8264	.8289	.8315	.8340	.8365	.8389
1.0	.8413	.8438	.8461	.8485	.8508	.8531	.8554	.8577	.8599	.8621
1.1	.8643	.8665	.8686	.8708	.8729	.8749	.8770	.8790	.8810	.8830
1.2	.8849	.8869	.8888	.8907	.8925	.8944	.8962	.8980	.8997	.9015
1.3	.9032	.9049	.9066	.9080	.9099	.9115	.9131	.9147	.9162	.9177
1.4	.9192	.9207	.9222	.9236	.9251	.9265	.9279	.9292	.9306	.9319
1.5	.9332	.9345	.9357	.9370	.9382	.9394	.9406	.9418	.9429	.9441
1.6	.9452	.9463	.9474	.9484	.9495	.9505	.9515	.9525	.9535	.9545
1.7	.9554	.9564	.9573	.9582	.9591	.9599	.9608	.9616	.9535	.9633
1.8	.9641	.9649	.9656	.9664	.9671	.9678	.9686	.9693	.9625	.9706
1.9	.9713	.9719	.9726	.9732	.9738	.9744	.9750	.9756	.9699	.9767
2.0	.9772	.9778	.9783	.9788	.9793	.9798	.9803	.9808	.9761	.9817
2.1	.9821	.9826	.9830	.9834	.9838	.9842	.9846	.9850	.9812	.9857
2.2	.9861	.9864	.9868	.9871	.9875	.9878	.9881	.9884	.9854	.9890
2.3	.9893	.9896	.9898	.9901	.9904	.9906	.9909	.9911	.9887	.9916
2.4	.9918	.9920	.9922	.9925	.9927	.9929	.9931	.9932	.9913	.9936
2.5	.9938	.9940	.9941	.9943	.9945	.9946	.9948	.9949	.9934	.9952
2.6	.9953	.9955	.9956	.9957	.9959	.9960	.9961	.9962	.9951	.9964

2.7	.9965	.9966	.9967	.9968	.9969	.9970	.9971	.9972	.9963	.9974
2.8	.9974	.9975	.9976	.9977	.9977	.9978	.9979	.9979	.9973	.9981
2.9	.9981	.9982	.9982	.9983	.9984	.9984	.9985	.9985	.9980	.9986
3.0	.9987	.9987	.9987	.9988	.9988	.9989	.9989	.9989	.9986	.9990
3.1	.9990	.9991	.9991	.9991	.9992	.9992	.9992	.9992	.9990	.9993
3.2	.9993	.9993	.9994	.9994	.9994	.9994	.9994	.9994	.9993	.9995
3.3	.9995	.9995	.9995	.9996	.9996	.9996	.9996	.9996	.9995	.9997
3.4	.9997	.9997	.9997	.9997	.9997	.9997	.9997	.9997	.9997	.9998

CO AND PO ATTAINMENT TABLE

Course outcomes (COs) for this course can be mapped with the programme outcomes (POs) after the completion of the course and a correlation can be made for the attainment of POs to analyze the gap. After proper analysis of the gap in the attainment of POs necessary measures can be taken to overcome the gaps.

Table for CO and PO attainment

Course Outcomes	Attainment of Programme Outcomes *(1- Weak Correlation; 2- Medium correlation; 3- Strong Correlation)*						
	PO-1	PO-2	PO-3	PO-4	PO-5	PO-6	PO-7
CO-1							
CO-2							
CO-3							
CO-4							
CO-5							
CO-6							

INDEX